ELECTRONICS SERVICING

ELECTRONICS SERVICING
Part 1 Radio, Television and Electronics Theory
City and Guilds of London Institute Course 224

Rhys Lewis, B.Sc.Tech., C.Eng., M.I.E.E.

Head of Department of Electronic & Radio Engineering,
Riversdale College of Technology, Liverpool

First published 1981 by
THE MACMILLAN PRESS LTD
London and Basingstoke
Associated companies in Delhi Dublin
Hong Kong Johannesburg Lagos Melbourne
New York Singapore and Tokyo

ISBN 0 333 28248 5

Typeset by Oxprint Ltd, Oxford
Printed in Hong Kong

Contents

Preface

This book is the first in a series designed to cover the City and Guilds of London Institute Course 224 Electronics Servicing, which is in the process of replacing Course 222 Radio, Television and Electronics Mechanics. The new course is divided into three parts, part I covering Electronic Systems and associated basic radio, television and electronic theory together with topics such as heat, light and sound, graphical communications, the binary system and calculations. Part II contains a common subject Core Studies and either Television and Audio Equipment or Industrial Equipment. At the time of writing, part III is in the process of being developed but in all probability will contain a series of specialist topics including Television Reception, Digital Applications of Television, AM/FM Reception and Audio Systems, Electronic Measurements and Control, Electronic Instruments and Testing and Microprocessor Computer Systems. This book covers the whole of the subject areas of part I, further books will cover part II Core Studies and the specialist areas in part II and possibly part III.

The book is divided into fourteen chapters under headings corresponding to the official syllabus but it should be noted that the layout is not necessarily the most appropriate learning or teaching order. The essence of both the new course and the one which preceded it is a 'systems' approach and accordingly the reader depending on his initial knowledge may care to start with signals and systems (chapters 7–10) and refer to the associated principles (chapters 1–6 and 14) as and when required. A selection of multiple choice questions covering the entire text is included at the end of the book.

I would like to thank the following firms for permission to use diagrams from their publications:
The Ever Ready Co. (Great Britain) Ltd., Whetstone, London;
Chloride Industrial Batteries Ltd., Swinton, Manchester;

Mr D. J. Hussey for his help with the material on Health and Safety and Miss Dawn Timmins for her invaluable assistance in the typing and preparation of the manuscript.

RHYS LEWIS
Liverpool
October 1980

1   Electrical supplies

Energy is defined as the ability to do work. There are many different kinds of energy—mechanical, electrical, chemical, heat and nuclear energy as well as electromagnetic energy and the energy of sound (which is a form of mechanical energy). All electronic systems in order to work must have a source of energy, in particular, electrical energy. Isaac Newton, the seventeenth-century English physicist, suggested that there is a fixed amount of energy available in the universe of different kinds which can be converted from one kind to another but never destroyed. Earlier this century Albert Einstein showed that energy can in fact be obtained from mass, a fact shown convincingly in nuclear reactions, which are the basis of electricity generation in nuclear power stations and, of course, nuclear bombs. Energy conversion is taking place all around us all the time. In the internal combustion engine the chemical energy of petrol is converted to the mechanical energy of movement of the car, in the electrical generator mechanical energy of whatever drives the generator is turned into electrical energy (see chapter 4) and in the electric fire or the electric lamp (the incandescent type) electrical energy is converted to heat or light respectively. In the case of the lamp both heat and light energy is available since the lamp gets warm.

The electrical energy necessary to make electronic systems work is normally derived from a generator source or from a battery source. The mains electricity supply in factories, shops, offices and homes is a generator source, the generators being situated in 'power stations' often some distance away and connected via overhead or underground cables. Mobile electronic systems in ships, aircraft or other vehicles often have their own generators with batteries providing additional or alternative electrical energy. Similarly, electronic systems sited in remote areas may have local generators with or without battery supplies.

Small portable electronic systems such as radios, recorders, calculators etc. normally use battery supplies and may have facilities for connection to mains supplies when these are available. In some special cases, for example, space vehicles, storage batteries are used which derive their energy from sunlight. In these cases sunlight is first converted to chemical energy which is then converted to electrical energy as required.

In this chapter we shall look in more detail at the types of electrical supply, how they work, what kind of supply they produce and where they are likely to be used. We shall also look at what an electrical supply produces in an electronic system and how one supply may be compared with another. First, however, as is usual in any branch of science or engineering we must look at measurement.

Electric Charge

When dry hair is combed with a plastic comb, the comb acquires the property of being able to attract small, light objects such as pieces of paper. The comb has acquired an *electric charge*. The method of charging a body, called charging by friction, has been known for many years, going back to the time of the Ancient Greeks. It is found that materials such as ebonite rubbed by fur or glass by silk acquire electric charges but that the charges are different in nature. Before anyone had any idea of what electric charge is or how it is transferred in this way these two kinds of charge were called positive and negative. It is found that bodies charged either positively or negatively will attract other uncharged bodies and each other. Bodies charged similarly, however, that is, both positively charged or both negatively charged, repel one another.

We now know that electric charge is carried within a material by a minute atomic particle called the *electron* and the charge it carries is the one that we call negative. (The electron was discovered after the charges were given their names.) When a body becomes negatively charged it is because it has gained electrons (transferred from one body to another), when a body becomes positively charged it has lost electrons. Uncharged bodies, being neither positive nor negative, are electrically neutral.

The electron is one of three fundamental small parts of the basic 'building block' of all materials, the *atom*. The other parts are the *proton* and the *neutron*. An atom, shown in diagram form in figure 1.1 is believed to be made up of a relatively heavy nucleus of protons and neutrons which is continually orbited by electrons moving in one of a number of elliptical paths, called *shells*. The proton is electrically positive, the neutron has no charge and an electrically neutral body has an equal number of positive protons and negative electrons. The orbiting electrons move at a distance from the nucleus which is determined by their energy. If given more energy from an external source they may leave their atoms (leaving the

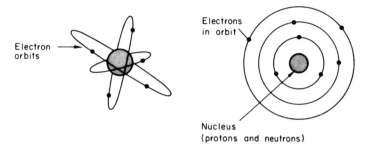

figure 1.1   Basic atomic structure

atoms positively charged since the protons now exceed the electrons in number) and may indeed leave the material altogether (hence, charging by friction). Similarly, an atom may gain additional electrons. When an atom loses or gains one or more electrons it becomes either negatively or positively charged and is then called an *ion*.

The unit of electric charge is called the *coulomb*, symbol C and is the charge carried by about 6.3×10^{18} electrons.

When electric charge moves continually, the flow of charge is called electric current. This is discussed further below.

Electromotive Force: Voltage

Earlier it was said that an electrical supply provides energy for an electrical or electronic system to work and that there are many kinds of energy. In the International System of Units all energy is measured in the same unit—the *joule*. (This is not true for many of the earlier systems of units which tended to make the study of engineering and science that much more confusing.) To gain a practical idea of the size of the joule imagine a one bar electric fire and the energy it releases (mainly heat, apart from the red glow). Such a fire usually uses 1000 joules every second and we can see that a joule is a fairly small unit. In discussing the energy given by an electrical supply to an electrical or electronic system we are concerned not only with the energy provided by the supply but also with what it is initially provided for—that is, we are concerned with the 'electrical' use to which the energy is put. In the first instance the energy given by the supply is given to electrically charged particles (such as electrons) and so we are interested both in energy and charge. The unit of electrical charge is the coulomb (the charge carried by 6.3×10^{18} electrons) so the usefulness of an electrical supply in providing energy for charged particles is measured in *joules per coulomb*. One joule per coulomb is given the special name of *volt* (after Alessandro Volta, the Italian scientist) and the abbreviation is V. As with other SI units multiples and sub-multiples may be used, for example, millivolt (one thousandth of a volt), kilovolt (one thousand volts) and so on. A table of these multiples and sub-multiples is given in chapter 14 (page 230).

A 100 V supply, then, provides 100 joules for each coulomb of charge (i.e. to 6.3×10^{18} electrons if the charge is carried by electrons), a 1000 V supply provides 1000 joules for each coulomb and so on.

The word *voltage*, commonly used, means the number of volts available or being used in a system and the voltage of a supply is given a special name—electromotive force, abbreviated e.m.f. In some ways this is an unfortunate choice of name since the quantity voltage of a supply is *not* a force (force in SI units is measured in newtons). However, it is a name which has been in use for many years, given, one suspects, at a time when electrical quantities were not as well understood as they are today.

Electric Current

Electric current is the movement of electrical charge carried usually (but not always) by electrons. If one coulomb of charge is being transferred from one point to another every second we say that the current flowing is one coulomb per second and this is given the special name the *ampere*, abbreviated A. Multiples and sub-multiple units of current include the microampere (one millionth of one ampere) and the milliampere (one thousandth of one ampere) abbreviated μA and mA respectively.

If the charge carriers move in one direction only the current is called direct current, abbreviated d.c. If the charge carriers periodically change their direction the current is called alternating current, abbreviated a.c.

Generator Sources

The principle of electricity generation using electromagnetism is dealt with in chapter 4. As stated earlier electromechanical generators may be used on site, close to the electronic systems they supply or may be situated elsewhere, the e.m.f. being connected to the system by means of the national distribution of overhead and underground cables. In the United Kingdom virtually all electricity generation is of the alternating type, the e.m.f. causing current to flow first in one direction and then the other at regular intervals called *cycles*. A graph of alternating voltage or current plotted against time is shown in figure 1.2.

The number of cycles generated per second is called the *frequency* of the supply, one cycle per second being called one hertz (after the German scientist who discovered electromagnetic radiation). The abbreviation for hertz is Hz and multiples and sub-multiples include kilohertz (kHz) megahertz (MHz) and gigahertz (GHz) being one thousand, one million and one thousand million cycles per second respectively.

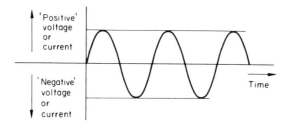

figure 1.2   Alternating current or voltage waveform

In the United Kingdom the generation frequency is standard at 50 Hz (in the USA and Canada it is 60 Hz) and this frequency is not allowed to vary by more than a few per cent. Alternating voltages and currents are used for a number of reasons particularly because of the ease of generation and distribution. It is found that the most efficient distribution is obtained at high voltages, generation taking place at lower voltages and the normal domestic and industrial supply being at still lower voltages for safety reasons. The levels of alternating voltages and currents can be changed very easily and efficiently using transformers (described in chapter 4). Generation and distribution of d.c. would be much more difficult and very much less efficient. In practice, generation takes place usually at 11–13 kV (sometimes at 3 kV) and this voltage is increased to 132 kV or as high as 400 kV for interconnection of overhead lines, supported by the familiar pylons scattered across the countryside, to which all generating stations may be connected as and when required. In this way when a generating station has more energy available than required for local use, this can be diverted to the Grid system. Similarly when demand locally exceeds available local supply energy may be drawn from the Grid. The normal working voltage for the Grid is 132 kV and for the so-called 'Supergrid' is 400 kV.

Domestic voltage is usually in the region of 230–250 V the reduction or 'stepping down' of the high national distribution voltage taking place at local 'sub-stations'. It is worth noting at this point that although the domestic voltages are much lower than those of generation or distribution they can be equally fatal if mishandled.

Various aspects of safety when dealing with electrical equipment and supplies are considered elsewhere in this book. Industrial voltages range from 440 V to 230 V, again the stepping down being carried out at factory sub-stations.

Almost all electronic systems require a d.c. supply and consequently supplies derived from the national 'mains' system must first be suitably altered before they can be used. This is the function of the 'power supply' unit within a mains connected electronic system.

Airborne or other vehicle electronic systems often derive their electricity supply from local generators carried with the system and the voltage and frequency of these supplies may vary from system to system. Aircraft and ships' systems often use a 400 Hz supply at various voltages and when servicing of such systems is necessary it is essential to first check on the initial supply type, voltage and frequency since these may determine to a large extent the characteristics of system components.

Battery Sources

A battery is a means of converting chemical energy to electrical energy. When connected to a conductive circuit it provides direct current flowing in one direction from the battery to the circuit along one lead and from the circuit back to the battery along a second lead. The direction of current flow in each lead, either from or to the battery remains the same and does not change as with alternating current. Conventionally, we say that current flows from the positive terminal of the battery to the circuit and from the circuit to the negative terminal of the battery. This is the direction of current flow, positive to negative, which was assumed when current was first discovered and before scientists knew of the existence of the electron.

In fact, electrons, which are usually (although not always) the charge carriers which form an electric current, flow from negative to positive and so electrons will move from the negative terminal of a battery to the circuit and from the circuit to the positive terminal of the battery. The reason for the use of 'conventional' current flow is that a number of useful 'rules of thumb' connecting current flow to magnetism and vice versa (in motors and generators, see chapter 4) were developed after the discovery of electric current but before the discovery of the electron and these rules are still used. In this book arrows showing current flow in circuit diagrams will indicate conventional current unless otherwise stated. In most American textbooks electron flow is used and consequently the 'rules of thumb' mentioned earlier have to be changed or abandoned.

A battery, then, is a chemical means of providing direct current to an electrical or electronic circuit. There are two main types of battery— *primary* and *secondary*. A primary battery is one which is of no further use and is disposed of once all the chemicals in the battery have been exhausted. These batteries, of the 'throw-away' variety, are used in radios, torches, clocks and watches, gas lighters, calculators, electrical and electronic toys, tape recorders and hearing aids, to name some of the common applications. Primary batteries cannot be recharged to any useful extent once the chemicals are exhausted and battery manufacturers warn that attempts to recharge such batteries can lead to a build-up of dangerous gas pressures inside them. Secondary batteries can be recharged by connection to a suitable source of electrical energy. These batteries, sometimes referred to as 'storage' batteries (although, strictly speaking, all batteries store energy), are used in cars and other vehicles for starting purposes, in electric vans and buses for traction (see figure 1.3) and as stand-by emergency supplies in the event of mains electricity failure.

figure 1.3   Electric traction

All batteries of either type consist of a number of *cells* connected together, each cell having two connecting points, one positive and one negative, and a number of chemicals contained within it. The cells may be connected so that the e.m.f. of each cell is added to its neighbour to give a total battery e.m.f. equal to the sum of the individual cell e.m.f.s or so that the battery e.m.f. is the same as the cell e.m.f. and the total current supplied from the battery when connected is equal to the sum of the currents provided by each cell. The first connection is called a *series* connection, the second is called a *parallel* connection. Batteries may also use, of course, a combination of series and parallel connections. Series and parallel connection of cells to form batteries is described in more detail in the next chapter.

Primary Cells

The modern primary cell, of which there are a number of varieties, is a development of the simple cell consisting of two rods, called *electrodes*, one zinc and one copper, immersed in a solution of sulphuric acid and water called the *electrolyte*.

As was stated earlier, atoms are electrically neutral but it is possible for an atom to gain or lose electrons without altering the nature of the original material and in this case the combination is called an *ion*. When the zinc and copper electrodes are immersed in the dilute acid the zinc begins to dissolve releasing negative zinc ions and the sulphuric acid begins to break up into positive hydrogen ions and negative sulphate ions. (Sulphate is a combination of sulphur and oxygen.) The zinc electrode becomes negative and the hydrogen ions move to the copper electrode making it electrically positive. Between the two electrodes, one now positive, the other negative, an e.m.f. of approximately 1 V is established.

If a simple cell is used to provide electric current it is able to do so for only a short period until one or other of three things occur. These are the complete dissolving of the zinc, the weakening of the acid rendering it inoperative and the formation of hydrogen at the positive (copper) electrode which eventually prevents current flow. This last occurrence is called 'polarisation'. Modern cells always contain a chemical to prevent polarisation, called a *depolariser*. The positive electrode is called the *anode* and the negative electrode is called the *cathode*.

Other cell arrangements have been developed which use different materials for anode and cathode and for the electrolyte and three of the more commonly used ones are briefly described below.

The Leclanché Cell

The original Leclanché cell invented by Georges Leclanché in 1866 consisted of a glass jar containing a positive electrode of carbon surrounded by manganese dioxide to act as a depolariser and, contained in a porous pot within the outer glass container, a negative zinc electrode and an electrolyte of ammonium chloride solution thickened by the addition of sand or sawdust. Although many variants have been successfully introduced, including using coke, graphite or carbon black in place of carbon and other materials for the depolariser, the basic materials of the Leclanché cell remains the same. See figures 1.4, 1.5 and 1.6, which show the construction of the modern version of the cell.

Leclanché dry cells are available in 'power pack' form for use in transistorised equipment where a relatively low current is required intermittently and in 'high power' form in equipment which requires higher current, for example, recorders, shavers and cine cameras. The single cell has an unloaded (open circuit) e.m.f. of 1.6 V falling to between 1.2 V and 1.4 V when fully discharged.

The Mercury Cell

Mercury cells have a zinc anode, a mercuric oxide cathode and the electrolyte is a potassium hydroxide solution. They are manufactured in

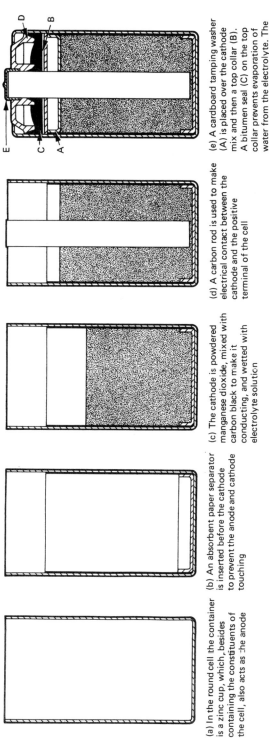

(a) In the round cell the container is a zinc cup, which, besides containing the constituents of the cell, also acts as the anode

(b) An absorbent paper separator is inserted before the cathode to prevent the anode and cathode touching

(c) The cathode is powdered manganese dioxide, mixed with carbon black to make it conducting, and wetted with electrolyte solution

(d) A carbon rod is used to make electrical contact between the cathode and the positive terminal of the cell

(e) A cardboard tamping washer (A) is placed over the cathode mix and then a top collar (B). A bitumen seal (C) on the top collar prevents evaporation of water from the electrolyte. The cell is closed at the top with a plastic top cover (D) and a metal top cap (E) and then fitted with a metal jacket (not shown)

figure 1.4   The construction of the Leclanché cell (courtesy of the Ever Ready Co.)

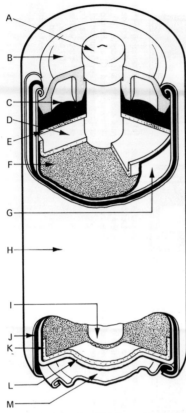

D. Top Washer. This functions as a spacer and is situated between the depolariser mix and the top collar.

E. Top Collar. This centralises the carbon rod and supports the bitumen sub-seal.

F. Depolariser. This is made from thoroughly mixed, high quality materials. It contains manganese dioxide to act as the electrode material and carbon black for conductivity. Ammonium chloride and zinc chloride are other necessary ingredients.

G. Paper Lining. This is an absorbent paper, impregnated with electrolyte, which acts as a separator.

H. Metal Jacket. This is crimped on to the outside of the cell and carries the printed design. This jacket resists bulging, breakage and leakage and holds all components firmly together.

I. Carbon Rod. The positive pole is a rod made of highly conductive carbon. It functions as a current collector and remains unaltered by the reactions occurring within the cell.

J. Paper Tube. This is made from three layers of paper bonded together by a waterproof adhesive.

K. Bottom Washer. This separates the depolariser from the zinc cup.

L. Zinc Cup. This consists of zinc metal, extruded to form a seamless cup. It holds all the other constituents making the article clean, compact and easily portable. The cup is also the anode and when the cell is discharged part of the cup is consumed to produce electrical energy.

M. Metal Bottom Cover. This is made of tin plate and is in contact with the bottom of the zinc cup. This gives an improved negative contact in the torch or other equipment, and seals off the cell to increase its leak resistance.

A. Metal Top Cap. This is provided with a pointed pip to secure the best possible electrical contact between cells.

B. Plastic Top Cover. This closes the cell and centralises the positive terminal.

C. Soft Bitumen Sub-Seal. A soft bitumen compound is applied to seal the cell.

figure 1.5   The round cell battery (courtesy of the Ever Ready Co.)

cylinder and flat pellet form, the latter being shown in figure 1.7. In this construction the anode is formed from high purity zinc powder pressed into pellets, the cathode is a mixture of mercuric oxide and graphite and the electrolyte is potassium hydroxide. The cell container is nickel plated steel, the plating being required to reduce the effects of corrosive reactions taking place between the electrolyte and the container.

Mercury cells can provide a steady voltage and relatively high currents over a long period and do not require 'resting' as in the case of the

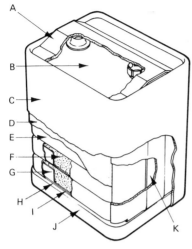

D. Wax Coating. This seals any capillary passages between cells and the atmosphere, so preventing the loss of moisture.

E. Plastic Cell Container. This plastic band holds together all the components of a single cell.

F. Depolariser. This is a flat cake containing a mixture of manganese dioxide as the electrode material and carbon black for conductivity. Ammonium chloride and zinc chloride are other necessary ingredients.

G. Paper Tray. This acts as a separator between the mix cake and the zinc electrode.

H. Carbon Coated Zinc Electrode. Known as a Duplex Electrode, this is a zinc plate to which is adhered a thin layer of highly conductive carbon which is impervious to electrolyte.

I. Electrolyte Impregnated Paper. This contains the electrolyte and is an additional separator between the mix cake and the zinc.

J. Bottom Plate. This plastic plate closes the bottom of the battery.

K. Conducting Strip. This makes contact with the negative zinc plate at the base of the stack and is connected to the negative socket at the other end.

A. Protector Card. This protects the terminals and is torn away before use.

B. Top Plate. This plastic plate carries the snap fastener connectors and closes the top of the battery.

C. Metal Jacket. This is crimped on to the outside of the battery and carries the printed design. This jacket helps to resist bulging, breakage and leakage and holds all components firmly together.

figure 1.6   The flat cell battery (courtesy of the Ever Ready Co.)

Leclanché cell. When a cell or, indeed, a power supply is able to provide steady voltage over a wider range of currents it is said to have good *regulation*. The regulation of the mercury cell is good even after storage for long periods at temperatures of 20 °C and above.

The Alkaline Manganese Cell

The electrolyte in the Leclanché cell is acid based. The mercury cell electrolyte has opposite characteristics to acid and is called *alkaline*. Another type of cell similar to the mercury cell using an alkaline electrolyte is the alkaline manganese cell. Like the mercury cell it is manufactured in cylinder and 'button' form, the cylindrical version being illustrated in figure 1.8. The essential difference between this cell and the mercury cell is the use of manganese dioxide and graphite for the cathode, the anode and electrolyte being powdered zinc and potassium hydroxide, respectively, as before.

The cell will provide heavy current for long periods and operates over a wide temperature range (−20 °C to 70 °C). It is not affected by long storage before use.

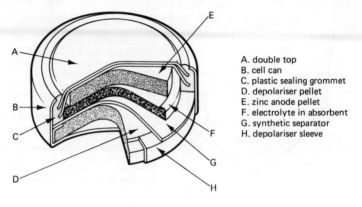

A. double top
B. cell can
C. plastic sealing grommet
D. depolariser pellet
E. zinc anode pellet
F. electrolyte in absorbent
G. synthetic separator
H. depolariser sleeve

figure 1.7   The mercury cell (courtesy of the Ever Ready Co.)

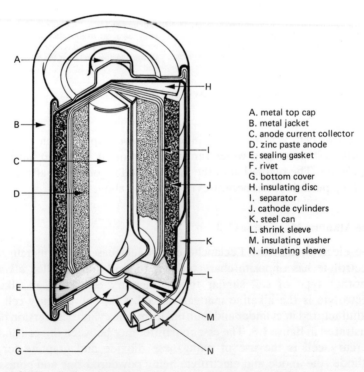

A. metal top cap
B. metal jacket
C. anode current collector
D. zinc paste anode
E. sealing gasket
F. rivet
G. bottom cover
H. insulating disc
I. separator
J. cathode cylinders
K. steel can
L. shrink sleeve
M. insulating washer
N. insulating sleeve

figure 1.8   The alkaline manganese cell (courtesy of the Ever Ready Co.)

Secondary Cells

Secondary batteries are made up of a number of secondary cells connected together. The important difference between a primary and secondary cell is that a secondary cell can be safely recharged once the cell voltage has fallen below a certain level. The charge–discharge process can be carried out many hundreds of times.

Secondary cells, like primary cells, consist of positive and negative electrodes immersed in an electrolyte which may be liquid or jelly. Liquid electrolyte or 'wet' cells have an opening at the top of the cell which is used for adding water to the cell and for obvious reasons the cell must be used in an upright position. The water should be pure and distilled water is normally used. Jelly electrolyte or 'dry' cells do not require 'topping up' with water and are factory sealed. They may be used in any position.

The commonest form of secondary cell consists of lead plates immersed in dilute sulphuric acid, the cell being called a lead-acid type (see figure 1.9). When fully charged the positive plates contain lead peroxide and the

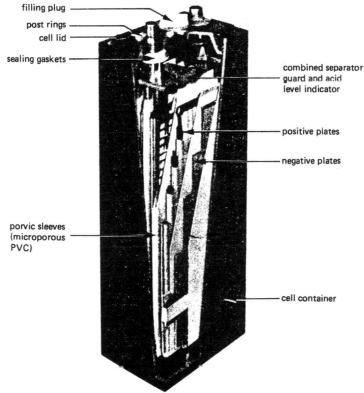

filling plug

post rings

cell lid

sealing gaskets

combined separator guard and acid level indicator

positive plates

negative plates

porvic sleeves (microporous PVC)

cell container

Lead-Acid Secondary Cell (Courtesy of Chloride Industrial Batteries Ltd)

figure 1.9a

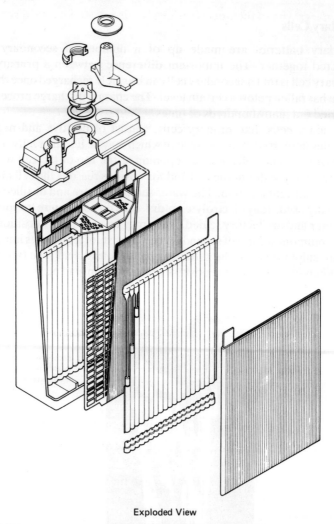

Exploded View

figure 1.9b

negative plates 'spongy lead'. On discharge the material of both plates combines with the acid to form lead sulphate, the acid becomes weaker and the cell voltage falls from 2.7 V to 1.8 V. The voltage should not be allowed to fall below this figure or 'sulphating' occurs in which both sets of plates become so heavily impregnated with lead sulphate that future use of the cell is impaired if not rendered impossible. An important characteristic of the wet electrolyte is its *specific gravity*, that is, the ratio between the weight of a given volume of the electrolyte and the weight of the same volume of water at the same temperature. When fully charged the specific gravity of the wet cell electrolyte is 1.27 and during discharge the specific gravity falls.

It should not be allowed to fall below 1.18. The easiest way of measuring specific gravity is by using a *hydrometer*, which consists of a graduated float contained in a glass syringe into which the electrolyte is drawn by means of a rubber bulb. The method of reading a hydrometer and its construction is shown in figure 1.10. It should be noted that the float is sometimes calibrated to read between 1100 and 1350 and in this case the reading shown is 1000 times the specific gravity of the electrolyte (that is the hydrometer range 1100 to 1350 gives a specific gravity of between 1.1 and 1.35).

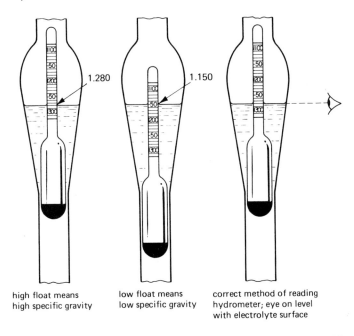

high float means   low float means   correct method of reading
high specific gravity   low specific gravity   hydrometer; eye on level
                                                    with electrolyte surface

figure 1.10   Reading a hydrometer (courtesy of Chloride Industrial Batteries Ltd.)

A second type of secondary cell using an alkaline electrolyte has nickel or its compounds as the positive plate and iron or cadmium for the negative plate, the electrolyte being potassium hydroxide. Nickel cadmium cells may be cylindrical or button in shape and in wet or dry form, commonly the latter (see figure 1.11). When fully charged the anode is nickelic oxide and the cathode cadmium, these materials changing to nickel oxide and cadmium hydroxide respectively during discharge. The electrolyte remains unchanged. The cell voltage falls from 1.3 V to 1.2 V to 1.0 V for the dry cell, there being no need to concern oneself with

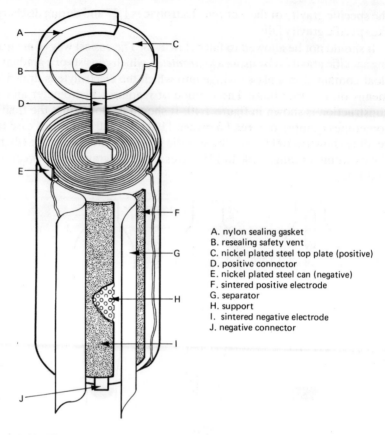

A. nylon sealing gasket
B. resealing safety vent
C. nickel plated steel top plate (positive)
D. positive connector
E. nickel plated steel can (negative)
F. sintered positive electrode
G. separator
H. support
I. sintered negative electrode
J. negative connector

figure 1.11   The nickel cadmium cell (courtesy of the Ever Ready Co.)

the electrolyte specific gravity for this type of construction. It should be noted that during charging the dry nickel–cadmium cell voltage rises to 1.45 V. Nickel–iron (NIFE) cells are of the wet variety and use nickel hydroxide and graphite for the anode and a mixture of iron and cadmium oxides for the cathode, the electrolyte being a solution of potassium hydroxide. The cell voltage is 1.7 V during charge, 1.2–1.33 V when fully charged and approximately 1.1 V when fully discharged. The specific gravity usually remains the same at 1.19 and should never be allowed to fall below this figure by more than 0.03.

All types of secondary cell should be maintained as specified by the manufacturer and with care a long useful life is easily obtained.

The capacity of a secondary cell or battery is the amount of charge available during discharge from fully charged to discharged. The coulomb is too small a unit of charge for this purpose so that the *ampere-hour* is used instead. The ampere-hour (symbol Ah) is the charge moved by a current of one ampere flowing for one hour (the coulomb is the charge moved by one ampere flowing for one second). Manufacturers quote a rate of charge (or discharge) for a secondary battery, usually given in terms of ampere-hours and time. Thus a 500 Ah, 10 h battery would give 50 A for 10 h. It is reasonable to assume that for *lower* currents the discharge time may be obtained by dividing the given capacity in ampere-hours by the current (thus a 500 Ah, 10 h battery would give 25 A for 20 h or 10 A for 50 h) but this is not true for higher currents since the rate of discharge affects the capacity, a higher rate resulting in lower capacity. The battery quoted would not give 100 A for 5 h since at this rate of discharge the capacity would be reduced from the quoted figure of 500 Ah. Normally, a secondary cell or battery is charged at the rate quoted, that is, a 10 h battery is charged for ten hours, but higher rates of charge can be tolerated especially in the case of nickel–cadmium cells.

2   Electrical circuits

The Conductive Circuit

An electronic system is made up of electronic units, each unit carrying out a particular job on the electronic signal as it passes through the system. Electronic units, in turn, consist of one or more *conductive circuits* and these are made up of electrical and electronic *components* connected together so that when a voltage source is applied to the circuit an electric current flows from the source, to and through the circuit and back to the source. If the source is a direct current source the current flow is in one direction (conventionally from positive to negative—but electrons flow in the opposite direction) and if the source is an alternating current source the current flow changes direction periodically. Components and materials used for connecting them are considered in more detail later; for the moment we must examine the basic ways of connecting components together. The component which will be used as an example is one with which most people are familiar, a small lamp similar to a torch bulb, which could be used for indicating purposes in a circuit.

Circuit Diagrams

Figure 2.1a shows a drawing of a simple 1.5 V cell connected to a lamp. Conventionally, current flows in the direction shown from the positive terminal of the cell through the lamp and back to the negative terminal of the cell. This figure is a drawing of the appearance of the lamp and the cell and is a *wiring diagram*. A wiring diagram shows the connecting wires in a circuit and exactly where they are joined. It is not necessary to always draw a detailed picture of the components and voltage source in a wiring diagram but the connection points as they appear are usually shown. Figure 2.1b shows another form of wiring diagram which indicates how

18

the top of the cell and lamp terminals might look.

A second form of diagram of a circuit is the *schematic diagram* which uses symbols for both voltage sources and components and may not necessarily show exact points of connection of wires. A schematic diagram of the cell and lamp circuit is shown in figure 2.1c. The point about the schematic diagram not always showing exact points of connection will be made clearer as the circuits become less simple later in this section.

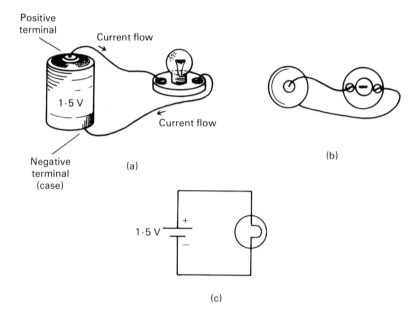

figure 2.1   (a) Cell and lamp; (b) top view; (c) schematic

Series Connected Loads

In any conductive circuit the interconnection of components which draws current from the voltage source is called the *load* on the source. The voltage source provides energy to drive electric current through the load. In the simple circuit of cell and lamp shown in figure 2.1 the lamp is the load on the 1.5 V cell. Now consider the circuits of figure 2.2. Figure 2.2a shows a load of two lamps and figure 2.2b shows a load of three lamps. In both circuits the lamps are connected to each other or to the cell such that the current from the cell must pass through each lamp in turn as it flows from the cell through the load and back to the cell. The connection is called a *series* connection. When two or more components are connected

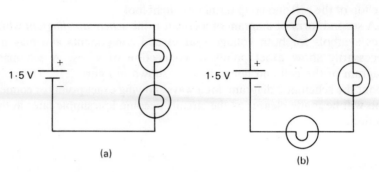

figure 2.2   (a) Two-lamp load; (b) three-lamp load

in series the same current passes through each component. This does not mean only the same value of current—it is possible in a more complicated circuit for the same value or level of current to flow in different parts of the circuit but they are not necessarily in series. A series connection means that the whole current in one component flows from that component into its series connected neighbour. In the circuits of figure 2.2 we are still using a 1.5 V cell as in the simple one lamp load circuit of figure 2.1. The same e.m.f. instead of having to drive current through one lamp is used to drive current through two lamps or three lamps. If the lamps are identical this means that they will each have available one half of the total e.m.f. if the load is two lamp or one third of the total e.m.f. if the load is three lamp. The lamps will therefore have less current flowing in them than in the one lamp load and will be less bright. How voltage is distributed in a series circuit is considered in more detail below and in chapter 3.

Series Connection of Cells

To increase the brightness of the lamps in the circuits shown in figure 2.2 to the level of the single lamp in the one lamp load, the voltage source may be increased by using series connection as shown in figure 2.3. Here we have two or three cells connected together in series so that their e.m.f.s aid one another. In the circuit of figure 2.3a two 1.5 V cells are connected with the positive terminal of one to the negative terminal of the other, leaving the remaining positive terminal of one cell and the remaining negative terminal of the other cell available for connection to the load. These two cells are now called a *battery* and, since they aid each other, the battery has an available e.m.f. of 1.5 V + 1.5 V, that is 3 V. Similarly, in the circuit of figure 2.3b, the brightness of the lamps is restored to that of the single lamp in the one lamp circuit by connecting three 1.5 V cells in *series-aiding* to give a 4.5 V battery. Sometimes a

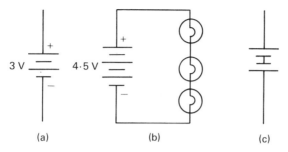

figure 2.3   Cells connected in series (a) and (b) aiding; (c) opposing

single cell is referred to as a battery but strictly speaking a battery means two or more cells connected together. It is of course possible to connect cells together so that they oppose one another as shown in figure 2.3c. Here the negative terminals of each cell are connected together leaving two positive terminals available for the load. A *series-opposing* connection like this achieves nothing so there is little point in making it. However, the principle should be known for in more complex circuits one may meet voltage sources each attempting to drive current in the opposite direction to the other and in this case the sources are then in opposition. An example is given in chapter 3.

Potential Difference

If a ball or other object is held above ground we say it has *potential energy*. Work is done and energy is used in raising it to its position and if it is then released it is pulled to the ground by gravity, accelerates and just before hitting the ground it will have *kinetic energy* equal to the potential energy prior to release (assuming no losses of energy due to air resistance etc.). Potential energy is energy by virtue of position relative to some fixed point (in this case, the ground) and kinetic energy is energy of movement. The potential energy of the ball or object is converted to other forms as the ball or object moves from the rest position to the ground.

In an electrical circuit we have a similar situation. Here a charged particle, an electron, is given energy from a voltage source and at a point just prior to entering the load it has a certain level of energy. On passing through the load this energy is converted to other forms—light and heat energy if the load is a lamp—and the potential energy of the particle on entering the load has changed to a new level by the time the particle leaves the load. The energy of the particle entering the load is called potential energy because it is energy by virtue of the particle's position with respect to a fixed point, in this case, the point of leaving the load. The difference in levels of the particle's potential energy entering and leaving the load is called the *potential difference* (p.d.) across the load. Since it is energy of an electrically charged particle it is measured in units of energy per unit of charge, that is, joules per coulomb or volts.

The analogy between electrical potential energy of electrons passing through a lamp and a falling object is shown in figure 2.4. Here the potential energy of a ball at two points above ground measured relative to the ground is shown as W_1 and W_2. For the electron the energy levels per unit charge are shown as V_1 and V_2 relative to some other point in the circuit.

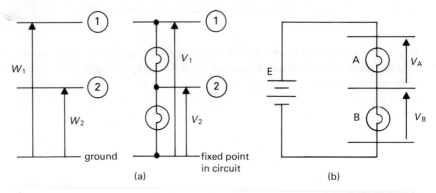

(a)　　　　　　　　　　　　　(b)

figure 2.4　Potential

In passing from position 1 to position 2 the potential energy of the ball changes from W_1 to W_2 and $W_1 - W_2$ units of energy are converted to other forms (namely kinetic—some will be lost in air resistance and the surrounding air will absorb the energy probably in the form of heat). The electron in moving from position 1 to position 2 has its level of potential energy changed from V_1 to V_2 and $V_1 - V_2$ units of energy per unit of charge are converted to other forms (heat and light since the load shown is a lamp). The potential difference across the lamp is $V_1 - V_2$ volts. In a complete circuit all the energy given by the source to each unit of charge is converted to other forms as the particles making up the unit (6.3×10^{18} electrons make up one coulomb of charge) move round the circuit. Thus, the e.m.f. applied to the circuit (the energy given by the source to each unit of charge) is equal to the *sum* of the potential differences around the circuit, since each p.d. represents an energy conversion per unit charge within the circuit component across which the p.d. is measured. As an example, in the circuit of figure 2.4b, which shows two lamps series connected to form a load supplied by a battery of e.m.f. E volts, the p.d. across lamp A is V_A and across lamp B is V_B. Thus, ignoring any p.d. existing across the connecting wires,

$$E = V_A + V_B$$

This is a mathematical way of saying that since the battery provides E joules/coulomb and in lamp A V_A joules/coulomb are converted to heat and light energy and in lamp B V_B joules/coulomb are converted to heat and light energy:

total energy per coulomb supplied (E) = total energy per coulomb converted $(V_A + V_B)$

In practice some energy is used in passing current through the connecting wires and there will be a small p.d. across them. It may usually be ignored. Note that the battery in this circuit is composed of two cells connected in series aiding. If they are identical cells the e.m.f. per cell will be $\frac{1}{2}E$ volts.

Parallel Connected Loads

When components are connected so that the same p.d. exists across them, they are said to be connected in *parallel*. By the same p.d. we do not mean merely a p.d. of the same value exists across the components. In a circuit containing two or more components it is quite possible that two or more of the components have p.d.s across them of the same value, the components themselves being in series. The same p.d. means that there exists across the parallel connection *one and only one* p.d. This is shown in figure 2.5 part (a) of which shows two parallel connected lamps, part (b) showing three parallel connected lamps. The p.d. across the lamps is shown as V volts.

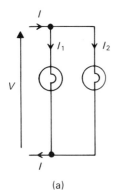
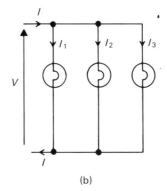

(a)   (b)

figure 2.5   Parallel connections (load)

In a series connected circuit the same current flows through each component, the p.d. across the components being separate (possibly equal in value but nevertheless separate). In a parallel connected circuit there is one p.d. but several currents, again possibly equal but nevertheless separate. In the circuit of figure 2.5a, for example, the p.d. is V volts and there are two currents shown as I_1 and I_2 amperes. These currents combine to form the total load current I. Current I flows into the load, divides into I_1 and I_2 which recombine to form current I leaving the load. I_1 and I_2 may be equal in value as, for example, if the lamps are identical but the lamps are nevertheless connected in parallel and not in series. Similarly, in the circuit shown in figure 2.5b, the load current I drawn from the voltage source divides into three component currents shown as I_1, I_2 and I_3, these currents recombining to form current I leaving the load and returning to the battery.

Voltage sources may be connected in parallel in the same way as components forming a load. If they have the same e.m.f. the total e.m.f. of the parallel connection is the same as that of a single cell. More current can be drawn from the voltage source, however, since each cell can contribute its maximum current and these add up to provide a load current as shown in figure 2.6. Here two cells of e.m.f. E V are connected in parallel each providing a current of I A. The overall voltage source now provides an e.m.f. E V, as for the single cell, but the load current is $2I$ A. If the parallel connected cells do not each have the same value of e.m.f. the resultant e.m.f. will have a value between the maximum e.m.f. and the minimum e.m.f. and this resultant value will depend on the internal characteristics of the cell. This is made clearer in chapter 3 after electrical resistance has been discussed.

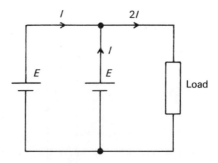

figure 2.6   Parallel connection of cells

Circuits consisting only of series or parallel connections are usually quite simple circuits. Most circuits which make up electronic units consist of combinations of series and parallel connected components. It is essential that the basic characteristics of the series connection and the parallel connection are known so that these connections may be recognised within a more complex circuit. Figure 2.7 shows a circuit having a composite load containing both kinds of connection. The voltage source

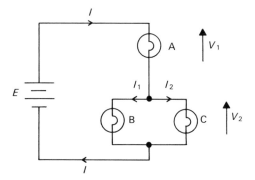

figure 2.7   Composite circuit

consists of two cells connected in series aiding and the load contains three lamps, shown as lamp A, B and C. Lamps B and C are connected in parallel and have a common p.d. shown as V_2. This parallel combination takes current I which divides into I_1 and I_2, these component currents recombining to form current I leaving the parallel connected lamps. Current I also flows through lamp A so that, by our definition of series connection, this lamp is in series with the parallel combination of lamps B and C.

It is not true to say that lamps A and B are in series or lamps A and C are in series since the same current does not flow in each. It cannot do so since the current in lamp A, shown as I, must divide as it enters the parallel combination of lamps B and C. The series connection is between lamp A and the parallel combination of lamps B and C together. The p.d. across lamp A is shown as V_1 and, from our earlier work, this p.d. added to p.d. V_2 across the parallel combination will be equal to the applied e.m.f. E, again ignoring any small potential difference across the wires connecting the load to the voltage source and the load components to each other.

Schematic and Wiring Diagrams

Earlier it was pointed out that circuit diagrams may be of the schematic type or the wiring type. Schematic diagrams use symbols for components and connections are shown to the nearest convenient point for the diagram to be electrically correct. Sometimes this means joining a wire apparently to the middle of another wire (see, for example, the connections to lamp 1 in the circuit of figure 2.5a); in practice separate wire connections are rarely ever made in this manner, terminal posts being used to ensure a good connection. (Printed circuits as described in chapter 12 join connecting paths to each other and sometimes the pattern of the copper track may correspond quite closely to the schematic diagram.)

The use of terminals posts in separately wired circuits means that often there is more than one way of wiring the circuit to produce the same electrical connections. Figure 2.7 shows a schematic diagram of a three lamp series-parallel load and in figure 2.8, we have two possible ways of wiring the circuit. In both parts of the figure, the terminals of the lamps are shown as circles as well as the positive and negative terminals of the battery. The lines now represent actual wires. In figure 2.8a, the bottom of lamp A is connected to the top of lamp B by one wire and to the top of lamp C by another wire.

In figure 2.8b the bottom of lamp A is again connected to the top of lamp B but on this occasion the top of lamp C is also connected to the top of lamp B. Since this in turn is connected to the bottom of lamp A, the top of lamp C is still electrically connected to the bottom of lamp A but this time via the lead from A to B. In both circuits the electrical connection is the same although the actual mechanical connections are made differently. Similarly, the negative terminal of the battery is connected to the bottom of lamp B in the circuit of part (a) of the figure, the bottom terminals of both lamps being joined, whereas in the circuit of part (b) of the figure the negative battery terminal is connected to the bottom of lamp C. Since again the bottom terminals are joined the same electrical connection is achieved, the actual 'joining points' being different. Both diagrams in figure 2.8 are possible wiring diagrams for the circuit shown in schematic form in figure 2.7.

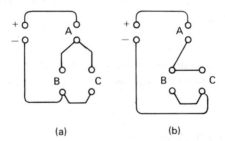

(a)   (b)

figure 2.8   Wiring the circuit of figure 2.7

Fuses and Switches

Any component used in an electrical circuit can carry only a limited current. Beyond this limit too much energy is absorbed by the component and damage will result. Usually this is as a result of overheating in the component. In order to prevent this happening and to protect components within a circuit some form of current limiting device is necessary. The simplest form of circuit protection is the *fuse*.

Basically a fuse consists of a wire made from material and of a physical size such that when a certain level of current is reached the wire melts and the fuse 'blows'. The fuse is connected in series with the load so that on melting the current path is broken and the load receives no further current. The current at which the fuse melts is *below* that which will damage the circuit. The melting current is called the fuse *rating* and this will be chosen to allow a safety factor between the rating and the minimum level of current likely to cause damage.

Thus if in a certain circuit a current of 0.5 A (500 mA) will cause damage, a fuse having a rating of say 250 mA or 300 mA may be used.

The fuse wire may be mounted between terminals in a *fuse holder*, one example of which is shown in figure 2.9a. This kind of holder is commonly used in house circuit protection, the body being made of ceramic material to prevent damage when the fuse wire melts. In the *cartridge fuse*, illustrated in figure 2.9b, the fuse wire is contained within a glass tube with metal end caps, the whole unit being replaced once the fuse has 'blown'. Larger cartridge fuses for heavy current use consist of ceramic tubes packed with special material to absorb the heat and other energy when the fuse wire melts. When a voltage source is first connected to a circuit which has been disconnected for some time larger than usual currents may flow for a short period, after which time the current settles down to a safe and acceptable level. An ordinary fuse may blow each time the voltage source is first applied in these circumstances if the initial current, called the *transient* current, is too high. In this case a special kind

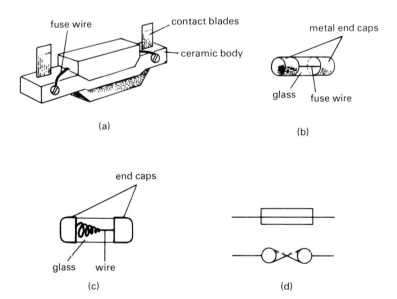

figure 2.9   Fuses. (a) Fuseholder; (b) cartridge fuse; (c) anti-surge fuse; (d) fuse symbols

of cartridge fuse, illustrated in figure 2.9c called a *surge protection* fuse may be used. This kind of fuse will allow larger than the rated current to flow for a very short period. If this higher level of current persists for any length of time, however, the fuse blows in the normal way. The symbols for fuses as used in schematic diagrams are shown in part (d) of figure 2.9.

Switches

A switch is a means of making and breaking a circuit connection, that is closing or opening a circuit. There is a tremendous variety of switches, operated manually or mechanically or electromechanically or electronically. A simple manual switch is shown in figure 2.10a, the ON marking indicating the switch position when the circuit is energised or 'live', the OFF position indicating when the circuit is not energised or 'dead'.

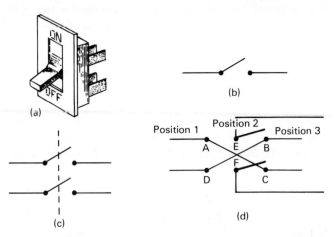

figure 2.10   Switches. (a) Manual switch; (b) single pole, single throw; (c) double pole, single throw; (d) double pole, double throw (reversing switch)

Some basic switch functions are shown in the remainder of figure 2.10. Part (b) shows the symbol for a switch which opens or closes a single connection with one movement. This type of switch is called single pole, single throw. Part (c) shows the symbol for a switch which opens or closes two connections simultaneously and is called double pole, single throw. Finally, the symbol for a switch which controls two connections but has two or three operating positions is shown in figure 2.10d. In position 1, E is connected to A, F to D. In position 2, E and F are not connected and in position 3, E is connected to B and F to C. If A and C are linked and B and D are linked the switch may be used for reversing as in figure 7.3a. The switch is called double pole, double throw. Position 2 may not always be available, in which case E moves from A to B and F moves from D to C.

3 Electrical resistance

Conductors and Insulators

If the same voltage is applied across pieces of different materials such as copper, iron, Nichrome (a mixture of nickel and chromium) and glass, which are of equal size and at the same temperature, the amounts of electric current which flow in each piece are different. The largest current flows in the copper, less current flows in the iron, less still in the Nichrome and the current in the glass will be so small as to be virtually undetectable. The reason for this is the different atomic structure of the materials. Copper atoms have outer electrons which are so loosely bound to their 'parent' atoms that only a small amount of energy per electron is required to free it. Glass, on the other hand, has atoms which contain tightly bound electrons and a great deal of energy per electron is required before electrons can move away and form an electric current. The other materials require varying amounts of energy per electron, Nichrome needing about ten times as much as iron. The same voltage (which is a measure of energy per unit charge and therefore energy per electron) will thus cause more current in the copper than in the iron, even less in Nichrome and least of all in glass. Copper and other similar materials such as silver, gold and aluminium are said to be good *conductors*, glass and materials such as rubber and plastics are good *insulators*. Materials between these two extremes may be called conductors or insulators depending on their electrical properties and there is a special class of materials called *semiconductors* which, strictly speaking, are neither conductors nor insulators.

The type of material used in an electrical circuit depends upon the use for which it is required. Good conductors are used to carry current with minimum reduction from one part of a circuit to another, insulators are used to electrically isolate parts of a circuit or circuit components from each other. When it is necessary to reduce the level of current within a circuit a conductor may be used, but one which offers greater opposition to current than, say, copper or aluminium.

Semiconductors such as silicon or germanium, because of special properties they have, are used in the manufacture of electronic components such as transistors and integrated circuits.

Resistance

Opposition to electric current is called electrical resistance or, within the subject of electricity and electronics, resistance since we are aware that it is electrical.

The resistance of anything carrying current, whether it is a part of a more complex circuit or a single component or piece of material on its own is determined by dividing the voltage (in volts) across the circuit or component by the current (in amperes) which is flowing. The result is expressed in *ohms*, the symbol for ohm being Ω (the Greek letter omega). An ohm is therefore a volt per ampere. As an example if a voltage of 1 V causes a current of 1 A, the resistance is 1 Ω. Similarly if 6 V causes 6 A to flow, the resistance is again 1 Ω (being 6 V divided by 6 A).

$$\text{Resistance } (\Omega) = \frac{\text{Voltage (V)}}{\text{Current (A)}}$$

This equation is an expression of *Ohm's Law*, which is named after the scientist Ohm, who discovered that the ratio of voltage to current for a material at constant temperature remains the same.

It is much easier to use symbols for quantities. Thus if we indicate resistance by R, voltage by V and current by I, we can write

$$R = \frac{V}{I}$$

and also, if both sides of this equation are multipled by I and the left- and right-hand sides of the equation interchanged

$$V = IR$$

Similarly dividing this equation throughout by R and writing I on the left-hand side

$$I = \frac{V}{R}$$

These three equations all say the same thing; which one is used depends upon which quantity is unknown and which quantities are known when a calculation is to be made. Examine the following examples of calculations.

Example 3.1

A voltage of 3.5 V is applied across each of three pieces of material in turn. The current which flows in each material is 0.5 A, 20 mA and 10 μA respectively. Calculate the resistance of each piece of material.

When $V = 3.5$ V, $I = 0.5$ A

$$R = \frac{3.5}{0.5} \qquad (R = \frac{V}{I})$$

$$= 7\Omega$$

When $V = 3.5$ V, $I = 20$ mA (that is $\frac{20}{10^3}$ A)

$$R = \frac{3.5}{20} \times 10^3$$

$$= 175\ \Omega$$

When $V = 3.5$ V, $I = 10\ \mu$A (that is $\frac{10}{10^6}$ A)

$$R = \frac{3.5}{10} \times 10^6$$

$$= 350\ 000\ \Omega$$

$$= 350\ k\Omega$$

Notice the use of the multiples, m, k and μ (milli, kilo and micro) meaning one thousandth, one thousand and one millionth respectively.

Example 3.2

Calculate the current flowing when 10 V is applied across materials of resistance 10 Ω, 20 kΩ and 5 MΩ respectively.

When $V = 10$ V, $R = 10\ \Omega$

$$I = \frac{10}{10} \qquad (I = \frac{V}{R})$$

When $V = 10$ V, $R = 20$ kΩ (that is 20×10^3 Ω)

$$I = \frac{10}{20 \times 10^3}$$

$$= \frac{10}{20} \text{ mA} \quad \text{(since } \frac{1}{10^3} \text{ A} = 1 \text{ mA)}$$

$$= 0.5 \text{ mA}$$

When $V = 10$ V, $R = 5$ MΩ (that is 5×10^6 Ω)

$$I = \frac{10}{5 \times 10^6}$$

$$= \frac{10}{5} \text{ }\mu\text{A} \quad \text{(since } \frac{1}{10^6} \text{ A} = 1 \text{ }\mu\text{A)}$$

$$= 2 \text{ }\mu\text{A}$$

Example 3.3

Calculate the voltage across a piece of material of resistance 10 Ω when it is carrying currents of 2 A, 20 mA and 50 μA respectively.

When $R = 10$ Ω, $I = 2$ A

$$V = 2 \times 10 \quad (V = IR)$$

$$= 20 \text{ V}$$

When $R = 10$ Ω, $I = 20$ mA (that is $\frac{20}{10^3}$ A)

$$V = \frac{20}{10^3} \times 10$$

$$= 0.2 \text{ V}$$

When $R = 10$ Ω, $I = 50$ μA (that is $\frac{50}{10^6}$ A)

$$V = \frac{50}{10^6} \times 10$$

$$= 500 \text{ }\mu\text{V}$$

$$\text{(or } 0.5 \text{ mV or } 0.0005 \text{ V)}$$

A useful method of remembering the three equations of Ohm's Law is the triangle shown in figure 3.1

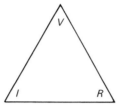

figure 3.1

The position of the letters V, I and R helps to remind us which quantity is divided or multiplied by one of the others to give the third.

For example: $I = \dfrac{V}{R}$

V is higher than R in the diagram so V is at the top of the right-hand side, R at the bottom.

Similarly $R = \dfrac{V}{I}$ since I is below V in the diagram

and $V = IR$ since both I and R are at the same level.

Factors Affecting Resistance

As we have seen the resistance of a piece of material depends to a great extent on the material itself, whether it is a conductor or insulator. In addition to this, however, resistance also depends on the physical size of the material, its length and area of cross-section—and on temperature.

To gain some idea why this should be so imagine yourself trying to get from one end of a busy street filled with shoppers to the other end. Clearly the longer the street the more effort is required, the wider the street and the more room there is to manoeuvre the easier it is. For an electron moving from one end of a piece of material to the other, the longer the piece the greater is the resistance. The comparison is over simplified but serves its purpose quite well. Resistance of a piece of material, then, increases with material length and is reduced if the material cross-sectional area is increased. The effect of temperature will be described later in the section. The equation which gives the relationship between resistance, length, cross-sectional area and type of material is:

$$\text{resistance} = \frac{\text{resistivity} \times \text{length}}{\text{cross-sectional area}}$$

and we can see that if length is increased the right-hand side of this equation is increased and so therefore is the resistance. If the cross-sectional area is increased, the right-hand side of the equation is decreased (since area is in the denominator of the right-hand side). The obvious question is—what is resistivity? Earlier it was said that resistance of any piece of material depends upon three things—material type, length and cross-sectional area. Since the last two are in the equation, resistivity must have something to do with material type.

And so it has. If we take a piece of material of length 1 m and cross-sectional area 1 m², we have

$$\text{resistance} = \frac{\text{resistivity} \times 1}{1}$$

that is resistance = resistivity

This tells us that the resistivity of a material is the resistance of a piece of the material of length 1 m and cross-sectional area 1 m². It is actually defined as the resistance between opposite faces of a cube of side 1 m (a 'unit cube'). A cube of side 1 m has a cross-sectional area of 1 m² and opposite faces are 1 m apart (figure 3.2):

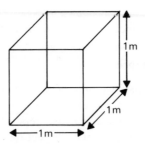

figure 3.2

The symbol for resistivity is ρ (pronounced 'ro').
Since temperature has an effect on resistance (and therefore resistivity) resistivities are always given at a particular temperature. The resistivity of some common materials at 20 °C is given below

Aluminium	0.0265	Glass	1×10^{18}
Copper	0.0172	Porcelain	2×10^{19}
Gold	0.0244		
Iron	0.1		
Magnesium	0.045		
Mercury	0.958		
Silver	0.0162		
Nichrome	1		

these figures being the resistance in millionths of ohms (microhms) between opposite faces of a cube of side 1 m. The unit of resistivity used here is the microhm-metre.

Looking at the table of values of resistivity we can easily see the difference between good conductors such as copper, silver and aluminium and good insulators such as glass or porcelain. Materials such as Nichrome lie between the two but much closer to the 'conductor' end of the range. Calculation of resistance of equal size pieces of different materials shows the effect of the material quite clearly.

Example 3.4

Using the figures for resistivity given, calculate the resistance of a piece of length 0.5 m and cross-sectional area 0.001 m² of each of the following materials; copper, aluminium, iron, Nichrome and glass.

For copper, $\rho = 0.0172$

$$\text{and using resistance} = \frac{\rho \times \text{length}}{\text{cross-sectional area}}$$

$$\text{resistance} = \frac{0.0172 \times 0.5}{0.001}$$

$$= 8.6 \ \mu\Omega \ \text{(microhms)}$$

For aluminium, $\rho = 0.0265$

$$\text{resistance} = \frac{0.0265 \times 0.5}{0.001}$$

$$= 13.25 \ \mu\Omega$$

For iron, $\rho = 0.1$

$$\text{resistance} = \frac{0.1 \times 0.5}{0.001}$$

$$= 50 \ \mu\Omega$$

For Nichrome, $\rho = 1$

$$\text{resistance} = \frac{1 \times 0.5}{0.001}$$

$$= 500 \ \mu\Omega$$

For glass, $\rho = 1 \times 10^{18}$

$$\text{resistance} = \frac{1 \times 10^{18} \times 0.5}{0.001}$$

$$= 500 \times 10^{18}\,\mu\Omega$$

(in words—five hundred million million million microhms).

The difference between copper or aluminium and glass is clearly emphasised. Notice that, as in all calculations, the units of quantities must be *compatible*. This means that if ρ is given in microhm-metres, the length must be written in metres and the cross-sectional area in square metres, the resistance then being in microhms.

To obtain an answer in ohms, divide the result by one million (10^6), giving in the above example the resistance of a piece of copper as $8.6/10^6$ ohms or $0.0000086\ \Omega$. The resistance of the piece of glass of the same physical size would be 500×10^{12} ohms or 500 000 000 000 000 Ω.

The Effects of Temperature on Resistance

As the temperature of a piece of material is increased a number of changes may occur. The length and cross-sectional area of the material may change as the piece expands (most materials increase in size as they are heated), the electrons within the material absorb energy making the free ones move faster and possibly releasing more electrons from the parent atoms. Each of these effects has a further effect on the resistance of the piece. Increased length means increased resistance, increased area means reduced resistance. More rapid movement in all directions of the electrons within the material may make it more difficult for conduction electrons to move through and thus effectively increase resistance. (Imagine the shoppers in the high street standing still and then moving about in all directions—it is clear which is easier for you to travel from one end of the street to the other.) Increased numbers of free electrons means more are available for current flow which may reduce the resistance. What happens overall to the resistance depends largely on the type of material, the overall effect on most conductors is an increase in resistance but this is not general for with some materials, especially those of a crystalline structure, the increase in free electrons overrides the other effects and resistance falls.

One of the most commonly used conductors is copper. The resistance of copper increases with increase in temperature and careful measurements show that a piece of copper of resistance 1 Ω at 0 °C has a resistance of 1.426 Ω at 100 °C, the resistance changing by the same amount for each degree rise in temperature. This amount may be obtained from the figures given, since for each ohm resistance at 0 °C the *change* in resistance for 100 °C is 0.426 Ω, and therefore the change per degree is

$$\frac{0.426}{100} \text{ or } 0.00426$$

This change in resistance per degree change in temperature which occurs for each ohm of resistance at 0 °C has a special name: the *temperature coefficient of resistance*, symbol α (alpha).

For copper $\alpha = 0.00426$ Ω per Ω per °C (written as 0.00426 Ω/Ω °C).

If we need to know the resistance of any particular piece of material at any temperature, given its resistance at another temperature and given its temperature coefficient of resistance we need an equation relating all these quantities. It is obtained as follows.

Suppose the resistance of a piece of material at 0 °C is R_0 Ω. Taking α to represent the temperature coefficient of resistance, each ohm will change by α ohms for each degree rise.

So R_0 Ω (the resistance at 0 °C) will change by αR_0 Ω for each degree rise. Now suppose we need to know the resistance at any temperature. Represent the temperature by t_1 and the resistance at t_1 by R_1 (t_1 being in °C and R_1 being in Ω). Then since the *change* in R_0 Ω for each °C rise is αR_0 Ω, for a rise of t_1 °C the change in R_0 will be $\alpha R_0 t_1$ Ω.

R_0 changes by $\alpha R_0 t_1$ Ω for a rise in temperature of t_1 °C.

The total resistance at t_1 = resistance at 0 °C + change in resistance

$$R_1 = R_0 + \alpha R_0 t_1$$
$$\underline{R_1 = R_0(1 + \alpha t_1)}$$

This is the required equation.

Example 3.5

The resistance of a piece of tungsten at 0 °C is 10 Ω. Its temperature coefficient of resistance is 0.005 Ω/Ω °C. Calculate the resistance of the tungsten at 100 °C.

Resistance at 100 °C = Resistance at 0 °C + change in resistance

Change in resistance = Resistance at 0 °C × new temperature × temperature coefficient of resistance

$$= 10 \times 100 \times 0.005$$

Resistance at 100 °C = 10 + 5
$$= 15 \ \Omega$$

or using the equation: $R_0 = 10, t_1 = 100, \alpha = 0.005$ the resistance at 100 °C

$$R_1 = 10\,(1 + 100 \times 0.005)$$
$$= 10\,(1.5)$$
$$= 15\,\Omega$$

If a piece of copper of resistance 1 Ω at 0 °C has its temperature progressively reduced the resistance will decrease, again by 0.00426 Ω for every °C reduction in temperature. Eventually its resistance will be zero and the number of °C reduction can be found as follows:

change in resistance = resistance at 0 °C × temperature coefficient of resistance × number of degrees change

The change in resistance is 1 Ω if the resistance at 0 °C is 1 Ω and the final resistance is to be zero.

So

$$\text{number of degrees change} = \frac{1}{1 \times \alpha}$$

$$= \frac{1}{0.00426}$$

$$= 234.5$$

and it is found that at −234.5 °C the copper would have zero resistance. Actually the change in resistance per °C is *not* constant below about − 50 °C so the figure is approximate.

However, the calculation indicates that very low resistance may be attained at extremely low temperatures and this is the principle behind 'superconductors'.

Sometimes we do not know the resistance of a material at 0 °C but at some other temperature. Again we can obtain an equation, but a slightly more complicated one than before. Let the resistance of a material at 0 °C be R_0 Ω then at any temperature t_1 °C, the resistance is given by

$$R_1 = R_0(1 + \alpha t_1)$$

(letting R_1 represent resistance at t_1 °C)

and at any other temperature t_2 °C, the resistance is given by

$$R_2 = R_0(1 + \alpha t_2)$$

(where R_2 represents the resistance at t_2 °C)

If the second equation is divided by the first we have

$$\frac{R_2}{R_1} = \frac{R_0(1 + \alpha t_2)}{R_0(1 + \alpha t_1)}$$

$$\frac{R_2}{R_1} = \frac{1 + \alpha t_2}{1 + \alpha t_1}$$

and we see that the resistance at 0 °C, R_0, disappears from the equation and we do not therefore need to know it.

$$R_2 = R_1 \frac{(1 + \alpha t_2)}{1 + \alpha t_1}$$

This equation connects R_2, the resistance at temperature t_2 °C with R_1, the resistance at temperature t_1 °C, α, the temperature coefficient of resistance and t_1 and t_2, the two temperatures.

This equation enables us to find the resistance of a piece of material at any temperature if we know its resistance at some other temperature.

Example 3.6

A coil of copper wire has a resistance of 200 Ω at 20 °C. Calculate the coil resistance at 50 °C if the temperature coefficient of resistance of copper is 0.0043 Ω/Ω °C.

Using the symbols of the equation we obtained above:

$$R_1 = 200 \ \Omega \quad , \quad t_1 = 20 \ °C$$
$$R_2 = ? \quad , \quad t_2 = 50 \ °C$$
$$\alpha = 0.0043 \ \Omega/\Omega \ °C$$

and we can find R_2 by using the equation

$$R_2 = R_1 \frac{1 + \alpha t_2}{1 + \alpha t_1}$$

$$= 200 \frac{1 + 0.0043 \times 50}{1 + 0.0043 \times 20}$$

$$= \frac{200 \times 1.215}{1.086}$$

$$= 226.76 \ \Omega$$

The resistance of the coil is 223.76 Ω at 50 °C.

We can also use the equation to find the temperature at which the resistance of a coil or other piece of material reaches a particular value—again assuming we know the resistance at some other temperature and the value of the temperature coefficient of resistance.

Example 3.7

The resistance of a coil of copper wire is 50 Ω at 15 °C. After current has been passed through the wire for a period of time, the resistance is again measured and found to be 60 Ω. Calculate the rise in temperature of the coil assuming the temperature coefficient of resistance of copper to be 0.0043 Ω/Ω °C.

The equation we must use is

$$R_2 = \frac{R_1(1 + \alpha t_2)}{1 + \alpha t_1}$$

where R_2 is the resistance in ohms at t_2 °C and R_1 is the resistance in ohms at t_1 °C, α being the temperature coefficient of resistance.

On this occasion we know R_1, R_2, t_1 and α and must find the value of t_2. To do this the equation must be *transposed*, that is rearranged to put t_2 on the left-hand side and the remaining (known) quantities on the right-hand side of the equation.

The method should be studied carefully.

1. Multiply both sides of the equation by $(1 + \alpha t_1)$

$$R_2(1 + \alpha t_1) = \frac{R_1(1 + \alpha t_2)(1 + \alpha t_1)}{(1 + \alpha t_1)}$$

$(1 + \alpha t_1)$ now cancels on the right-hand side to give

$$R_2(1 + \alpha t_1) = R_1(1 + \alpha t_2)$$

2. Multiply out the right-hand side to give

$$R_2(1 + \alpha t_1) = R_1 + \alpha R_1 t_2$$

3. Subtract R_1 from both sides

$$R_2(1 + \alpha t_1) - R_1 = R_1 - R_1 + \alpha R_1 t_2$$

to give $R_2(1 + \alpha t_1) - R_1 = \alpha R_1 t_2$

4. Divide both sides by αR_1

$$\frac{R_2(1 + \alpha t_1) - R_1}{\alpha R_1} = \frac{\alpha R_1 t_2}{\alpha R_1}$$

and R_1 cancels on the right-hand side to give

$$\frac{R_2(1 + \alpha t_1) - R_1}{\alpha R_1} = t_2$$

which, when t_2 is written on the left-hand side gives the required equation:

$$t_2 = \frac{R_2(1 + \alpha t_1) - R_1}{\alpha R_1}$$

It is not advisable to try and remember this equation. It is better if it is obtained when required. The essential thing to remember when re-arranging any equation is that whatever is done to one side of an equation in the way of multiplication, division, subtraction etc., *must* be done to the other. If this is not done the truth of the equation (that the left-hand side is equal to the right-hand side) is altered and the equation is then no longer valid.

Using the rearranged equation, we have, since

$R_2 = 60\ \Omega$, $R_1 = 50\ \Omega$, $t_1 = 15\ °C$ and $\alpha = 0.0043\ \Omega/\Omega\ °C$

$$t_2 = \frac{60(1 + 0.0043 \times 15) - 50}{0.0043 \times 50}$$

$$= \frac{60(1 + 0.0645) - 50}{0.215}$$

$$= \frac{60 \times 1.0645 - 50}{0.215}$$

$$= \frac{63.87 - 50}{0.215}$$

$$= 64.51\ °C$$

and the rise in temperature is $64.51 - 15$, that is $49.51\ °C$.

This kind of calculation is useful as a means of determining temperature of a component or other piece of material by measuring resistance.

Power

Power is the rate of providing or using energy. It is measured in units of energy per unit of time. The SI unit of energy is the joule and of time is the second, and the unit of power is given the special name *watt*, symbol W. As with all units we may use multiples and sub-multiples.

kilowatt	kW	1 000 watts
megawatt	MW	1 000 000 watts
milliwatt	mW	1/1000 watt
microwatt	μW	1/1000 000 watt

In an electrical circuit energy is given to electrically charged particles (usually electrons) by the voltage source and if current flows the energy is then given up by the charge-carriers as they travel round the circuit. The energy may be given up in the form of heat, as in an electric fire, or light, as in an electric light bulb, or in some other form.

Suppose the voltage across a circuit or part of a circuit at any particular time is 5 V and the current flowing is 2 A. This tells us that at the time we are looking at the circuit there are 2 coulombs per second flowing and each coulomb has 5 joules of energy. (Remember that one ampere is one coulomb per second and one volt is one joule per coulomb.)

If each coulomb has 5 joules of energy and 2 coulombs are flowing per second then the energy per second of the charge-carriers is

$$5 \, \frac{\text{joules}}{\text{coulomb}} \times 2 \, \frac{\text{coulombs}}{\text{second}} = 10 \, \frac{\text{joules}}{\text{second}}$$

This tells us that the power of this circuit or part of the circuit is 10 joules/second, or 10 W, at the time we are looking at the circuit.

Power in Resistive Circuits

We know that for a circuit or part of circuit in which the voltage is V volts and current I amperes, then power is given by

$$P = VI \text{ watts}$$

where P represents power.

If the circuit or circuit-part has resistance $R \, \Omega$ we also know from Ohm's Law that

$$V = IR$$

or, alternatively,

$$I = \frac{V}{R} \quad \text{or} \quad R = \frac{V}{I}$$

We can put these relationships in the power equation to give new equations relating power, voltage, current and resistance.

1. Replace V in $P = VI$ using $V = IR$

$$P = (IR)I$$

$$\underline{P = I^2R \text{ watts}}$$

2. Replace I in $P = VI$ using $I = \dfrac{V}{R}$

$$P = V \; \frac{V}{R}$$

$$P = \frac{V^2}{R} \; \text{watts}$$

We now have equations giving power in terms of voltage and current ($P = VI$), power in terms of current and resistance ($P = I^2R$) and power in terms of voltage and resistance ($P = V^2/R$). In a calculation we choose which one to use by looking carefully at the information given.

Whether this power is being given to the circuit or being given up by the circuit depends upon what the 5 V represents. If it is the voltage of an e.m.f. (a voltage source), then the source is delivering to the circuit 10 W at the time at which the voltage is 5 V and the current 2 A. If the 5 V is the potential difference across the circuit or part of the circuit we are looking at, the 10 W is being given back by the circuit at this time.

In general, if at a particular instant of time the voltage across a circuit or component is v volts and the current is i amps then the power at that instant is

$$vi \; \text{watts}$$

If we wish to determine power over a period of time we may still multiply together voltage and current provided that they are constant over that time. In direct current circuits in which a steady state has been reached, i.e. voltage and current levels have settled down to constant values, this is true, so that for a steady state d.c. circuit, if the voltage is V volts and the current I amperes, the power is VI watts.

In alternating current circuits both current and voltage levels are changing all the time and although the voltage $\times$ current equation for power still applies for power at a particular instant of time (instantaneous power), over a period of time other factors have to be considered. We will not go further into a.c. circuits at the moment; the remainder of this section will be concerned with methods of calculations which are valid for steady state d.c. circuits and for instantaneous values of power in a.c. circuits.

Example 3.8

Calculate the power given to a circuit of resistance 15 Ω, when the e.m.f is 5 V d.c.

Use

$$\text{power} = \text{voltage}^2/\text{resistance}$$

$$\text{power} = \frac{5^2}{15}$$

$$= 1.67 \; \text{W}$$

Example 3.9

Calculate the power absorbed by a resistor of value 1 kΩ when a circuit of 10 mA is flowing through it.

Use   power = current² × resistance

$$\text{power} = \left(\frac{10}{1000}\right)^2 \times 1000$$

(note that 10 mA is $\dfrac{10}{1000}$ A, 1 kΩ is 1000 Ω)

$$\text{power} = \frac{100 \times 1000}{1000 \times 1000}$$

$$= \frac{100}{1000}\ \text{W}$$

$$= 100\ \text{mW}$$

(since 1 mW = $\dfrac{1}{1000}$ W)

Example 3.10

The maximum safe power that can be absorbed by a certain part of a d.c. circuit is 0.5 W. Its resistance is 100 Ω. Calculate (a) the maximum value of current that can be allowed to flow, (b) the maximum voltage which can be applied to the circuit part.

This problem gives us the values of power and resistance but not voltage and current. These we have to find.

(a) Use $P = I^2R$ since we know the values of P and R and wish to find the value of I

$$0.5 = I^2 \times 100$$

$$I^2 = 0.005 \text{ (dividing both sides by 100)}$$

$$I = 0.0707 \text{ A}$$

$$= \frac{70.7}{1000} \text{ A (multiplying numerator and denominator by 1000)}$$

$$I = \underline{70.7 \text{ mA}}$$

(b) Use $P = V^2/R$ since we know the values of P and R and wish to find the value of V

$$0.5 = \frac{V^2}{100}$$

$V^2 = 50$ (multiplying both sides by 100)

$V = 7.07$ V

For (b) we could have obtained V from $P = VI$ since P is given and I has been calculated. This means, however, that we are using a figure for I which has been calculated rather than a value of R that has been given. To avoid any possibility of error in part (b) as a result of any error which may have occurred in part (a), the method of using $P = V^2/R$ is advisable.

We may use $P = VI$ as a check for our answers, however:

using $I = 0.0707$ A and $V = 7.07$ V

$$\text{power} = 0.070 \times 7.07$$

$$= 0.5 \text{ W as given}$$

More Complex Circuits: Resistors in Series and in Parallel

When components are connected so that the same current flows through each—not just the same value but the same current—they are said to be connected in *series*.

Figure 3.3a shows two resistors connected in series. The same current,

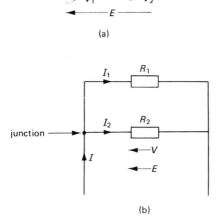

(a)

(b)

Resistors in Series and in Parallel

figure 3.3

I, flows through each resistor. The current is caused by an e.m.f., shown as E, and a p.d. exists across each resistor, its value depending on the value of the resistor since this determines how much energy is required from the charge carriers as they move through the resistor. The p.d. across resistor R_1 is shown as V_1 and across resistor R_2 is shown as V_2. The total energy per unit charge supplied by the source is E joules/coulomb, the energy per unit charge used in passing through R_1 is V_1 joules/coulomb and the energy per unit charge used in passing through R_2 is V_2 joules/coulomb. The total energy per unit charge used in passing through the series connection of R_1 and R_2 is $V_1 + V_2$. Thus, $E = V_1 + V_2$.

For resistor R_1, voltage = current × resistance, that is

$$V_1 = I \times R_1$$

For resistor R_2

$$V_2 = I \times R_2$$

and

$$V_1 + V_2 = IR_1 + IR_2$$
$$= I(R_1 + R_2)$$

so that

$$E = I(R_1 + R_2)$$

For the circuit as a whole

total voltage (e.m.f.) = current × total resistance

and by comparison we see that the effective total resistance of this connection is the sum of the resistance of the two components connected in series.

The total resistance of components connected in series is the sum of the individual resistances.

When components are connected so that the same voltage is applied across each, they are said to be in *parallel*. The resistors shown in figure 3.3b are in parallel. Here, the e.m.f., E, is applied across R_1 and R_2 whose ends are connected together so that the same voltage exists across each. The potential difference across each is the same and is equal to the e.m.f. Each resistor carries a current determined by the value of the resistor; the higher the resistance of the resistor the lower the current, since voltage/resistance is equal to current. The currents are shown as I_1 and I_2, the total current being shown as I.

Since there is no alternative path for the charge carriers to take, the number of carriers approaching the junction shown is equal to the number leaving the junction. Current means the amount of charge flowing per second, and, since each carrier has the same charge, the current depends on the number of charge carriers. Hence we can say that if the number of carriers approaching the junction equals the number of carriers leaving the junction, then the current flowing towards the junction, I, is equal to the sum of the currents I_1 and I_2 leaving the junction.

$$I = I_1 + I_2$$

For resistor R_1, voltage = current × resistance or current = voltage/resistance, thus

$$I_1 = \frac{V}{R_1}$$

For resistor R_2

$$I_2 = \frac{V}{R_2}$$

$$\text{total current} = I_1 + I_2$$

$$= \frac{V}{R_1} + \frac{V}{R_2}$$

$$= V \frac{1}{R_1} + \frac{1}{R_2}$$

For the circuit as a whole

$$\text{total current} = \frac{\text{total voltage}}{\text{total resistance}}$$

$$= \frac{V}{\text{total resistance}}$$

and by comparison we see that the total resistance of the circuit as a whole is equal to the reciprocal of $1/R_1 + 1/R_2$. Or, denoting total resistance by R

$$\frac{V}{R} = \frac{V}{R_1} + \frac{V}{R_2}$$

and

$$\frac{1}{R} = \frac{1}{R_1} + \frac{1}{R_2}$$

dividing throughout by V.

The total resistance of resistors connected in parallel is the reciprocal of the sum of the reciprocals of the individual resistances.

As is often the case, expressing a principle in words rather than symbols makes the principle appear more complex than is necessary. A useful quantity we can use here is *conductance*, which is the reciprocal of resistance. Conductance is a measure of the ease with which an electric current can be established in a circuit or component, whereas resistance is a measure of the difficulty experienced when establishing current.

$$\text{Conductance} = \frac{1}{\text{resistance}}$$

$$= \frac{\text{current}}{\text{voltage}}$$

The symbol normally used is G and the unit is the *siemens*, symbol S. (Note that the singular of the unit is siemens, not siemen.) Replacing $1/R$ by G, $1/R_1$ by G_1 and $1/R_2$ by G_2 in the equation above we have

$$G = G_1 + G_2$$

and we can say that *the total conductance of resistors connected in parallel is the sum of the individual conductances.*

The following examples should be studied carefully. They demonstrate a number of basic electrical principles.

Example 3.11

When 5 V is applied across a certain resistor, a current of 2 A flows. Calculate the resistance.

Solution

$$\text{Resistance} = \frac{\text{voltage}}{\text{current}}$$

$$= \frac{5}{2}$$

$$= 2.5 \ \Omega$$

Example 3.12

Find the resistance and conductance of a resistor if a current of 30 mA flows when a voltage of 100 V is applied across it.

Solution

A current of 30 mA (30 milliamperes) means 30/1000 amperes

$$\text{Resistance} = \frac{\text{voltage}}{\text{current}}$$

$$= \frac{100}{30/1000}$$

$$= 100 \times \frac{1000}{30}$$

$$= 3333.3 \ \Omega$$

$$= 3.333 \ k\Omega \ \text{(kilohms)}$$

$$\text{Conductance} = \frac{1}{\text{resistance}}$$

$$= \frac{1}{3333.3}$$

$$= 0.0003 \ S$$

Example 3.13

The following is a table of observations of voltages, currents and resistances. Fill in the gaps.

Voltage	Current	Resistance
10 V	3 A	?
?	20 mA	65 Ω
87 mV	?	1.2 kΩ
55 kV	90 μA	?

Line one

$$\text{resistance} = \frac{10}{3}$$

$$= 3.33 \Omega$$

Line two

$$\text{voltage} = \frac{20}{1000} \times 65$$

$$= 1.3 \text{ V}$$

Line three

$$\text{current} = \frac{87/1000}{1200}$$

$$= \frac{87}{1200 \times 1000}$$

$$= 0.000\ 072\ 5$$

$$= 72.5\ \mu\text{A}$$

Line four

$$\text{resistance} = \frac{55\ 000}{90/1000\ 000}$$

$$= \frac{55\ 000\ 000\ 000}{90}$$

$$= 611\ 111\ 111$$

$$= 611.1\ \text{M}\Omega$$

(In calculations involving submultiples such as this a better idea is to use powers of ten once you are able to.)

Example 3.14

Find the resistance of R in the circuit in figure 3.4.

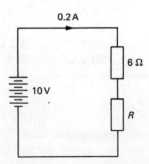

figure 3.4

Solution

$$\text{Total resistance of circuit} = \frac{\text{e.m.f.}}{\text{current}}$$

$$= \frac{10}{0.2}$$

$$= 50 \, \Omega$$

but

$$\text{total resistance} = 6 + R$$

(resistors in series) hence

$$50 = 6 + R$$

and

$$R = 44 \, \Omega$$

Example 3.15

Calculate the value of the total resistance when three resistors of resistance 15 Ω, 27 Ω and 33 Ω are connected in parallel.

Solution

Let R be the required total resistance. Then

$$\frac{1}{R} = \frac{1}{15} + \frac{1}{27} + \frac{1}{33}$$

$$= 0.134$$

$$R = 7.46 \, \Omega$$

Example 3.16

Find the value of R in the circuit shown in figure 3.5.

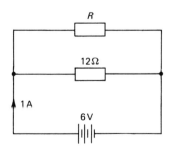

figure 3.5

Solution

$$\text{Total circuit resistance} = \frac{\text{voltage}}{\text{current}}$$

$$= \frac{6}{1}$$

$$= 6\,\Omega$$

and

$$\frac{1}{\text{circuit resistance}} = \frac{1}{12} + \frac{1}{R}$$

that is

$$\frac{1}{6} = \frac{1}{12} + \frac{1}{R}$$

$$\frac{1}{R} = \frac{1}{6} - \frac{1}{12}$$

$$= \frac{1}{12}$$

hence

$$R = 12\,\Omega$$

Example 3.17

In the circuit shown in figure 3.6 find (a) total circuit resistance and conductance, (b) the p.d. across each resistor and (c) the value of I_1, I_2 and I.

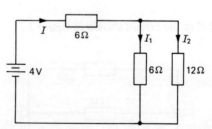

figure 3.6

Solution

(a) The total resistance is the resistance of a 6 Ω resistor in series with the parallel combination of two resistors, one 6 Ω and one 12 Ω. Let R be the resistance of the parallel combination.

$$\frac{1}{R} = \frac{1}{6} + \frac{1}{12}$$

$$= \frac{12 + 6}{72}$$

$$= \frac{1}{4}$$

hence

$$R = 4\,\Omega$$

Hence

$$\text{total resistance} = 4 + 6 = 10\,\Omega$$

$$\text{conductance} = \frac{1}{10} = 0.1\,\text{siemens}$$

(b) The current I is the total current, and total current from source is found from total voltage/total resistance, hence

$$I = \frac{4}{10}$$

$$= 0.4\,\text{A}$$

The p.d. across the 6 Ω resistor is $6I$, that is, $6 \times 0.4 = 2.4\,\text{V}$, then

$$\text{p.d. across parallel combination} = \text{e.m.f.} - 2.4\,\text{V}$$

$$= 4 - 2.4$$

$$= 1.6\,\text{V}$$

since the total p.d. is equal to the total e.m.f.

(c) The value of I has already been found above.

$$I_1 = \frac{\text{p.d. across 6 Ω}}{6}$$

$$= \frac{1.6}{6}$$

$$= 0.267\,\text{A}$$

$$I_2 = I - I_1$$
$$= 0.4 - 0.267$$
$$= 0.133 \text{ A}$$

(Alternatively, $I_2 = $ (p.d. across 12 Ω/12) = 1.6/12 = 0.133 A as before).

Example 3.18

Find the total resistance and conductance of the circuit shown in figure 3.7.

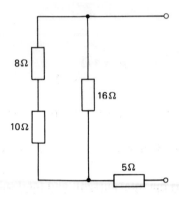

figure 3.7

Solution

8 Ω in series with 10 Ω is equivalent to 18 Ω. This 18 Ω resistance is in parallel with 16 Ω and so the equivalent resistance R is given by

$$\frac{1}{R} = \frac{1}{18} + \frac{1}{16}$$

Hence

$$R = 8.47 \, \Omega$$

This 8.47 Ω resistance is in series with the 5 Ω resistor and the total equivalent resistance of the whole circuit is

$$8.47 + 5 = 13.47 \, \Omega$$

$$\text{Total conductance} = \frac{1}{13.47} = 0.074 \, \text{S}$$

4 Magnetism

The effects of magnetism, or, to be more precise, electromagnetism, have been known for a long time. In both early Chinese and early Greek civilisations it was known that pieces of certain materials, when allowed to hang freely, always pointed in a certain direction; the fact was used as an aid to the navigation of ships. Electricity and magnetism are in fact related, magnetism being caused by the movement of electric charge (as we shall discuss shortly) but the connection between the two was not suggested before the seventeenth century and was not completely verified until the twentieth century.

Magnetism is the property of setting up a force which attracts certain materials. Not all materials are affected by a magnetic force; those which are affected are called *ferromagnetic* materials, the most common being iron, nickel and cobalt. Materials containing these elements also exhibit magnetic properties. Materials containing iron are called *ferrous* materials, the stem of the word ferromagnetic and ferrous being the same (it is derived from the Latin word for iron). The materials used by the Greeks as an early form of ship's compass, called *lodestone*, is now known to be a ferrous material. The magnetic force which causes the lodestone always to set in the same direction is exerted by the Earth itself. Anything which sets up a magnetic force is called a *magnet*. A piece of ferromagnetic material is not necessarily a magnet, that is, it does not necessarily *exert* a magnetic force, although it will always react to a magnetic force exerted by another magnet. Whether or not a ferromagnetic material behaves as a magnet depends on the material's state of *magnetisation*. This will be further discussed later.

If a piece of ferromagnetic material which is a magnet (usually called a bar magnet if that is its shape) is freely suspended it will always set itself in a certain direction, one end pointing north, the other south. The end pointing north is called the north-seeking or north pole of the magnet, the other end being the south-seeking or south pole of the magnet.

Magnetic Fields: Magnetic Flux

The area surrounding a magnet in which magnetic forces can be felt is called a *magnetic field*. One of the most famous nineteenth-century scientists, Michael Faraday, suggested that to represent a magnetic field in diagram form we should use lines to show the direction of action of the force set up due to magnetism. These lines he called *lines of force, lines of magnetic flux* or just magnetic flux. Originally the number of lines drawn showed the relative strength of the force within the magnetic field, nowadays although we still use the idea of magnetic flux, the strength of the magnetic field is measured differently.

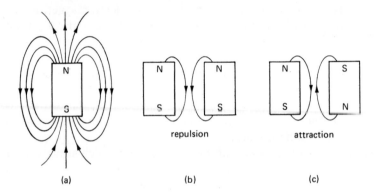

figure 4.1

Lines of force are drawn conventionally from the north pole of the magnet to the south pole and the *field pattern* of a simple bar magnet is shown in figure 4.1a. This pattern can easily be demonstrated by placing a sheet of paper over a bar magnet and gently shaking iron filings over the paper. The filings tend to behave as very tiny magnets and set themselves in appropriate directions in the magnetic field to form a pattern of lines like that shown in figure 4.1a. If two bar magnets are placed side by side there is an interaction between their respective fields and a force is set up between the magnets. If like poles are side by side (north next to north and south next to south) the force is one of repulsion (see figure 4.1b). If unlike poles are adjacent the force is one of attraction (see figure 4.1c). In general when two fields interact, like fields (lines of force in the same direction) repel and unlike fields (lines of force in opposite directions) attract. The force set up between magnetic fields is used in the electric motor to produce motion.

The Magnetic Field Set up by a Conductor

When an electric current flows through a conductor a magnetic field is set up around the conductor. The situation is shown in figure 4.2a. The lines of flux are circular and surround the conductor as shown in the figure. Looking at the conductor end on, if the current is flowing away from the observer, the lines of flux have a clockwise direction and if the current is flowing towards the observer the lines of flux have an anticlockwise direction. In diagrams current flowing away is shown by a cross and current flowing towards is shown by a dot when viewing end on, as shown in this figure. The current direction given is that of conventional current flow (positive to negative); electrons move in the opposite direction. A useful aid to memory here is the corkscrew—turning a corkscrew clock-

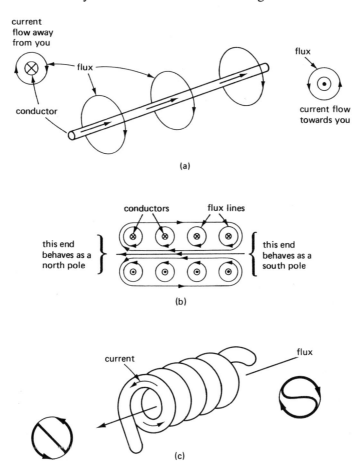

figure 4.2

wise moves the point away from the person turning, turning it anti-clockwise moves it back towards the person turning. The direction of turn corresponds to the flux direction and the respective direction of move-ment of the point corresponds to the direction of current flow producing the flux.

The strength of the magnetic field produced by an electric current flowing in a conductor may be considerably increased by winding the conductor in the form of a coil as shown in figures 4.2b and c. Figure 4.2b shows a cross-section of the coil in a plane through the coil centre parallel to the coil axis (perpendicular to the plane in which any one turn lies). The flux produced by each conductor *adds* to the flux produced by its neighbour giving a resultant increased flux as shown. Figure 4.2c shows a useful way of remembering directions in this case: looking end on at the coil, when current flows clockwise the flux direction is away from the observer and when current flows anticlockwise the flux direction is towards the observer. When the flux direction is away from the observer into the coil end, then that end behaves as the south pole of a magnet, the other end behaving as the north pole. When arrowheads are placed on the letters S and N as shown and the letters are drawn within a circle as shown, the arrows show the direction of current flow in the coil. Alter-natively, flux and current directions of figure 4.2a are current and flux directions respectively in figure 4.2c.

The Process of Magnetisation

As stated above, when an electric current flows a magnetic field is set up in the vicinity of the current, that is a magnetic force is established which will have an effect on ferromagnetic materials near by. Basic atomic theory indicates that all materials are made up of atoms, which in turn consist of a nucleus orbited by electrons. Now an electric current is a movement of electrically charged particles and thus the electrons orbiting within an atom constitute an electric current. A number of orbits pro-duces the equivalent of a number of minute currents and each current sets up a magnetic field. In the majority of materials the small magnetic fields within the material act in various directions such that the resultant field is zero. In ferromagnetic materials, however, it is believed that some of the internal fields act together to produce a number of what are called 'magnetic domains'. These domains may be considered to behave like a large number of small bar magnets. In an unmagnetised ferromagnetic material the domains are situated at random throughout the material and no resultant field is produced.

This is illustrated in diagram form in figure 4.3 which shows the domains as arrows pointing in many different directions (figure 4.3a). If the material is placed in an external field the domains begin to line up (figure 4.3b) and a small resultant field is set up. The material is now partially magnetised and is capable of exerting its own field. Eventually as the external field is increased in strength all the domains are aligned (figure 4.3c) and the material is fully magnetised.

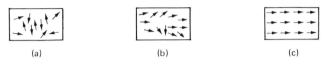

(a)   (b)   (c)

figure 4.3

If the external field is removed the domains may return to the state shown in figure 4.3b that is, the material is still partially magnetised, or may remain fully magnetised, depending on the material. Soft iron, for example, loses most of the magnetism when the field is removed but steel retains most of its magnetism. Steel and metals with similar properties (steel with nickel or certain alloys) are used for making permanent magnets. A permanent magnet retains its magnetism and is capable of setting up its own field if the magnetising field is taken away. A permanent magnet may then be used to make other magnets (the usual method uses a coil and electric current to set up the external field). Non-magnetic materials which do not have these domains are not affected by external fields at all.

Magnetic Circuit Quantities

An electrically conductive circuit is a collection of pieces of material joined together such that a source of voltage (an electromotive force or e.m.f.) causes an electric current to flow through the material in a continuous path. The opposition to current flow, measured as voltage/current, is called electrical resistance. Many electromagnetic machines and other devices (for example, relays and circuit breakers) may be considered to be or to have a *magnetic circuit* and it is a useful exercise to compare quantities in the conductive circuit with those in the magnetic circuit.

Figure 4.4a shows a simple magnetic circuit. It consists of a ferromagnetic material in the form of a square with the centre missing, so that it is made up of four pieces or *limbs*. The left-hand limb carries a coil of wire through which an electric current flows. (The coil is part of a separate electrically conductive circuit.) Using the theory already given lines of force or flux may be drawn through the coil, the top end of the coil behaving like the north pole of a bar magnet. Because of the nature of the circuit—it is made up of ferromagnetic material—the magnetic field is stronger within the circuit than outside it. Thus flux lines need only be drawn within the limbs as shown. The quantities in this magnetic circuit comparable to e.m.f. and current in the conductive circuit are the quantity which causes the field to be set up and some quantity which represents the field itself.

What are these quantities? First of all let us consider what sets up the magnetic field contained within the limbs. Clearly the electric current in the coil helps to establish the field, since without it the only magnetism is that of the material making up the limbs, if this material has magnetism (although it is ferromagnetic it may be in a state of non-magnetisation as explained earlier). However, it is not only the current which matters, it is also the coil, or more specifically, the number of turns on the coil. The more turns there are for any particular current value the stronger is the resulting magnetic field.

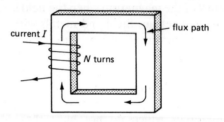

(a) Simple magnetic circuit

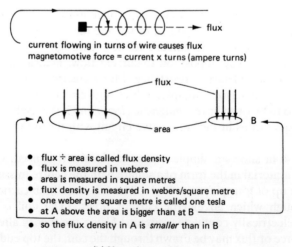

current flowing in turns of wire causes flux
magnetomotive force = current x turns (ampere turns)

- flux ÷ area is called flux density
- flux is measured in webers
- area is measured in square metres
- flux density is measured in webers/square metre
- one weber per square metre is called one tesla
- at A above the area is bigger than at B
- so the flux density in A is *smaller* than in B

(b) Magnetic quantities

figure 4.4

The name given to the quantity which sets up a magnetic field is *magnetomotive force*, abbreviated to m.m.f. As with the term electromotive force, the name is unfortunate since the quantity is *not* a force, although in this case, it is responsible for the setting up of conditions in which a force may be experienced. In the International System of units the quantity m.m.f. is measured by multiplying together the current and the number of turns; the unit is therefore the *ampere-turn*. This a 100-turn coil carrying a current of 0.5 A, for example, has an m.m.f. of 100×0.5 or 50 ampere-turns.

To find a magnetic quantity comparable to electric current we take the idea of flux one stage further than being merely a means of diagrammatically representing magnetic fields and imagine the flux as being set up by the m.m.f. Although, strictly speaking, flux is not a physical quantity like electric current, it is nevertheless a very useful concept which we can use to compare magnetic fields. Flux, like current, may be considered to be the 'effect' in the circuit due to the appropriate 'cause' (m.m.f. or e.m.f. respectively). Flux is given a unit (we can measure it in terms of its effects), the unit being defined in terms of other units (discussed later). The special name given to the unit of flux is the *weber*, symbol Wb.

Thus, summarising these ideas, magnetomotive force causes magnetic flux in a similar way to electromotive force causing electric current. m.m.f. is measured in ampere-turns and magnetic flux in webers. The method of defining a weber (to give an idea of the size of the unit) is discussed later in the chapter.

Magnetic Flux Density

If we assume that magnetic flux is a measurable quantity we can go one stage further in our efforts to compare magnetic fields and consider how this flux is distributed. If the same flux is distributed in two pieces of similar material, one being of larger cross-section than the other, it seems clear that, speaking loosely, the magnetic effect in the smaller piece is greater than in the larger piece, since the same flux is concentrated in a smaller area. This leads us to the idea of *flux density*, which is the magnetic flux (webers) divided by the cross-sectional area through which the flux acts (area in square metres). The unit of flux density is thus the weber per square metre, one weber per square metre being given the special name *tesla*, symbol T (see figure 4.4b).

Force Exerted on a Conductor Placed in a Magnetic Field

Earlier it was stated that in general when magnetic fields are close to one another, a force is set up between them, either of attraction or of repulsion (figure 4.1). Now a conductor carrying an electric current has its own magnetic field, so if it is placed in an external field, a force is exerted on the conductor and, if it is able to do so, it will move. The resulting field pattern will be considered later. For the moment consider the size of this force and what it depends on. First, it will depend on the flux density of the external field—the flux density rather than the flux alone because the denser the field the greater will be the force. Secondly, it will depend on the magnitude of the current carried by the conductor— because the field set up by the conductor is determined in size by the value of the current causing it and the force between fields depends in turn on the relative sizes of the fields concerned. Finally, the force depends on the length of the conductor: the longer the conductor the greater the force because more of the conductor field reacts with the applied external field.

If SI units are used, the force on a conductor placed in a magnetic field is given by

force in newtons = flux density in teslas × length of conductor in metres × current in amperes

or, using symbols

$$F = BlI$$

where F is the force, B the flux density, l the length and I the current.

Example 4.1

Calculate the current flowing in a wire placed in a magnetic field of flux density 0.1 T if the force per metre length of the conductor is 0.5 N.

Solution

$$\text{Force} = \text{flux density} \times \text{length} \times \text{current}$$

so that

$$\text{current} = \frac{\text{force}}{\text{flux density} \times \text{length}}$$

and, substituting values

$$\text{current} = \frac{0.5}{0.1 \times 1}$$

$$= 5 \text{ A}$$

Example 4.2

A conductor of length 0.2 m carrying a current of 50 A is placed in a magnetic field of flux density 0.6 T. Calculate the force on the conductor assuming it is at right angles to the field.

Solution

$$\text{Force} = 0.6 \times 0.2 \times 50$$

$$= 6\,\text{N}$$

Note that for the above formula to be correct the conductor is at right angles to the field of which the flux density is used in the equation. If the conductor is at any other angle, then the flux density is that of the field *component* which is at right angles to the conductor. (A magnetic field may be considered to have components in much the same way as a force or other vector quantity.)

The field pattern when a current-carrying conductor is placed in a magnetic field is illustrated in figure 4.5. The resultant field due to like fields repelling and unlike fields attracting is shown in the figure.

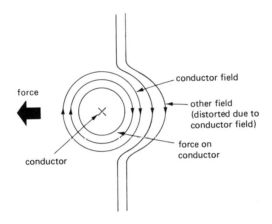

figure 4.5

It is useful to regard lines of force as being 'elastic' in their nature in that they attempt to take up the shortest length. If the lines were elastic, a force would be exerted in the direction shown, pushing the conductor to the left. The principle is used in electric motors and in indicating instruments using moving coils.

The Motor Principle

If a coil is pivoted about its axis so that it is free to rotate, and a magnetic field is set up so that the coil conductors are at right angles to the field flux as shown in figure 4.6a, a force is exerted on both conductors of the coil as shown in figure 4.6b. The coil will tend to rotate about its axis. This is the basic principle of motors and moving-coil indicating instruments. If the

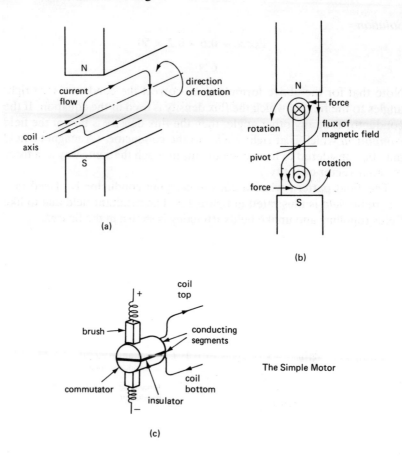

figure 4.6

principle is being used in a motor some means must be found of allowing the electric current to flow into and out of the coil as it rotates and the current must always flow in the appropriate directions shown in the figure, that is, into the paper in the top conductor and out of the paper in the bottom conductor in figure 4.6b. The simplest way of doing this is to use a *commutator*, as shown in figure 4.6c. The positive side of the supply is connected to the top brush and the negative side to the bottom brush; the brushes rub against the two segments connected to the conductors in turn as the coil rotates. In this way current always flows into the top conductor and out of the bottom conductor. In practice the main field is supplied by permanent magnets only in small machines and in permanent-magnet-moving-coil (p.m.m.c.) meters; in larger machines and in *dynamometer*-type meters the main field is supplied by a coil.

Some Other Applications of Electromagnetism

Figure 4.7 shows some applications of electromagnetism using a magnetic field set up by a current flowing in a coil. Figure 4.7a shows a simple bell; when current flows the coil magnetises its core and the clapper arm is attracted to the coil core. The coil is called a *solenoid* in this application. When the arm moves across it opens the switch, cutting off the current supply, and if the material is of the correct type, demagnetisation occurs and the arm springs back to its rest position. When it does so the switch is closed, current flows again and the cycle is repeated. Each time the arm moves across, the clapper strikes the gong giving the familiar bell sound.

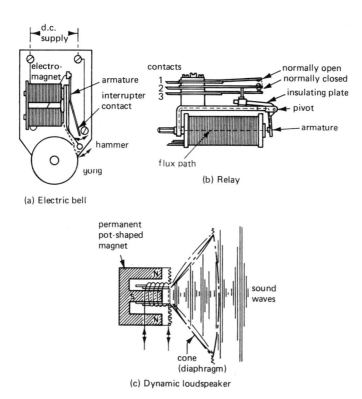

(a) Electric bell

(b) Relay

(c) Dynamic loudspeaker

figure 4.7

Figure 4.7b shows an electromagnetic relay which operates switches; when the coil current flows the arm moves in and the side arm outwards, opening or closing the switch as shown. In the 'normally open' arrangement the side arm pushes the contact arms together, in the 'normally closed' arrangement the side arm goes through a hole in one of the contact arms and the contact arms are pushed apart when the relay operates. The flux path is shown by a dotted line. The relay coil may be operated by a very low voltage and the switches used to operate high-voltage circuits; thus a high-voltage circuit may be operated quite safely and without danger to the operator.

Figure 4.7c shows a moving-coil loudspeaker which changes *alternating current signals* into sound. An alternating current signal is an electric current which flows alternately backwards and forwards through the coil, the change in current magnitude being used to carry information. For example an alternating current signal is obtained from a *microphone* which turns sound into electrical impulses. The loudspeaker carries out the opposite process to the microphone. The speaker coil is free to move along the middle part of the E-shaped core. The reaction between the fixed magnets of the core and the fluctuating field of the coil causes a varying force to be set up and this in turn moves the cone in sympathy with the force and thus with the signal causing the force.

In all these applications a suitable material is used as the core of the electromagnetic coil arrangement.

How the Ampere is Defined

As we have seen, the unit of electric current, the ampere, is the current flowing when one coulomb of electric charge moves per second. However, the ampere is not *defined* in this way. It is defined using the force that is set up when two conductors carrying current are placed a certain distance apart. We have seen above that when a conductor carrying current is placed in a magnetic field there is a force set up which is exerted on the conductor. If the magnetic field referred to is also set up by a conductor carrying current we have a situation in which a force exists between two conductors carrying current.

The ampere is that current which, when flowing in each of two infinitely long parallel conductors, situated in a vacuum and separated 1 m between centres, produces a force between these conductors equal to 2×10^{-7} N/m length.

Considering this formal definition further, 'infinitely long' can be taken to mean very long (there is a slight change in field patterns for short conductors), 'in vacuum' is necessary because the strength of the field or, to be precise, the flux density of the field, depends to some extent on the medium between the conductors (being slightly different in air, for example, than in a vacuum) and, finally, the rather strange figure 2×10^{-7} (meaning 2/10 000 000) is chosen deliberately so that relationships between other units depending on the ampere are as simple as possible.

Electromagnetic Induction

Many effects in science are reversible. Faraday, knowing that an electric current produced a magnetic field, wondered if the reverse were true. Did a magnetic field produce an electric current? He found that under certain circumstances it did. If a conductor is *moved* through a magnetic field or if a magnetic field is *moved* past a conductor (by moving a magnet past a conductor, for example) or if both field and conductor are stationary and the strength of the magnetic field is *changed*, then an e.m.f. is set up across the conductor, *provided that the magnetic field does not act in a direction parallel to the conductor.* If the conductor is part of a closed conductive circuit then current flows. The important necessary condition is that the magnetic flux *linking* the conductor should change. In the two cases involving movement, of either field or conductor, the magnetic field itself may be constant but, provided there is motion, the flux *linking* the conductor is changing from one instant to the next and the e.m.f. appears. The process is called *electromagnetic induction.*

If a coil is used to generate an e.m.f., then a number of conductors is involved and the *flux linkage* is defined as the product of flux and number of turns on the coil. Faraday discovered that *the value of the induced e.m.f. is proportional to the rate of change with time of the flux linkage.* If S I units are used this statement of proportionality becomes an equation as follows

induced e.m.f. = rate of change of flux linkage
            = rate of change of (number of turns × flux)

and if the number of turns is constant and therefore does not change

induced e.m.f. = number of turns × rate of change of flux

This is a statement of Faraday's law of electromagnetic induction.

Another scientist, named Lenz, discovered that the voltage induced always acts in a direction so as to oppose what is causing it to be set up. This is illustrated in figure 4.8. Figure 4.8a shows a conductor moving into a magnetic field acting vertically downwards (the north pole or effective north pole of whatever is causing the field being at the top of the diagram). If current flows in the conductor due to the induced voltage across the conductor it will flow in a direction such that the conductor field set up by the current opposes the main magnetic field into which the conductor is moving. Now like fields repel, so the conductor field must act in the direction shown in figure 4.8b. Using the corkscrew rule mentioned earlier, current must flow into the paper, that is, away from the observer, as indicated by the cross in the conductor. If current does *not* flow (because the conductor is not part of a complete circuit) the induced e.m.f. will act in a direction so as to cause the current direction

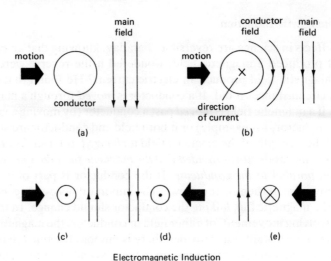

Electromagnetic Induction

figure 4.8

given *if the circuit was complete*, that is, the end of the conductor nearest to us is positive. (Remember that the direction of current flow is that of conventional current, which flows from positive to negative.) The remainder of the figure shows current directions for other field directions and other directions of motion of the conductor. This method of determining the direction of action of an induced voltage is superior to most other methods.

Because of the opposing nature of the induced voltage it is called a *back e.m.f.*

Example 4.3

The flux density of a magnetic field contained in a material of uniform cross-sectional area 0.09 m² changes from 0.1 to 0.25 T in 0.08 s. If a 50-turn coil were situated in a magnetic field changing in this way, the field being at right angles to the conductor, determine the average voltage induced across the coil.

Solution

$$\text{Flux density} = \frac{\text{flux}}{\text{area}}$$

flux = flux density × area

= 0.1 × 0.09 Wb when flux density is 0.1 T

and   flux = 0.25 × 0.09 Wb

when flux density is 0.25 T

$$\text{flux change} = (0.25 \times 0.09) - (0.1 \times 0.09)$$
$$= 0.0135 \text{ Wb}$$

which occurs in 0.08 s.

$$\text{Average rate of change of flux} = \frac{0.0135}{0.08}$$
$$= 0.169 \text{ Wb/s}$$

$$\text{Average e.m.f. induced} = 50 \times 0.169 \text{ (number of turns} \times \text{average rate of change of flux)}$$
$$= 8.44 \text{ V}$$

Example 4.4

A straight conductor 0.7 m long is moved with constant velocity at right angles both to its length and to a uniform magnetic field. Given that the e.m.f. induced in the conductor is 5 V and the velocity is 20 m/s calculate the flux density of the magnetic field.

Solution

Suppose the required flux density is represented by B (the usual symbol), then

$$\text{flux} = B \times \text{area over which flux acts}$$

The conductor length 0.7 m moves 20 m every second, the situation being shown in figure 4.9. The area over which the flux causing the induction acts is thus 20×0.7, that is 14 m². Thus

$$\text{flux} = 14\,B$$

and this flux is cut every second, so that

$$\text{rate of change of flux} = 14\,B \text{ Wb/s}$$

(the velocity is constant). This is a single conductor, the number of turns therefore is one, so using

$$\text{e.m.f.} = \text{number of turns} \times \text{rate of change of flux with time}$$
$$5 = 14\,B$$

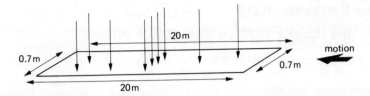

figure 4.9

and

$$B = \frac{5}{14}$$

$$= 0.357 \text{ T}$$

The flux density is 0.357 T.

Note that in these examples care was taken to say that the magnetic field linking the conductor was acting at right angles to the conductor. If the field is *parallel* to the conductor then electromagnetic induction *does not* take place.

Generation of an Alternating e.m.f.

Electromagnetic induction is used in the generation of voltage for domestic and industrial use. A simple generator is shown in figure 4.10. The similarity between this figure and figure 4.6, which shows a motor, is marked, the point being that generation is the motor principle applied in

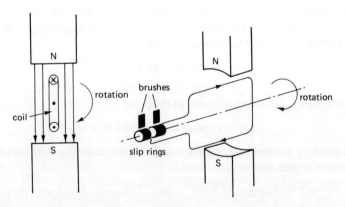

figure 4.10

reverse. With a motor, the current and field are supplied and current results (voltage actually, but current if the circuit is closed). If the induced voltage across the coil is picked *off* using brushes and the *slip-rings* shown, the voltage between brushes will reverse polarity as the conductor moves first under the north pole then under the south pole. The magnitude also changes because the *effective* flux cutting the conductor at the right angle required is changing as rotation takes place, i.e. only part of the full flux is cutting the conductor.

This results in a voltage waveform which is *sinusoidal*.

Sinusoidal Variation

A sinusoidally varying voltage-time or current-time graph is shown in figure 4.11. The voltage or current (produced by a sinusoidally varying voltage) rises from zero to a positive maximum then falls to zero, to rise to a maximum value in the opposite direction and return to zero before the process begins again. It should be noted that voltage actually changes polarity and current reverses its direction of flow as the variation moves through zero. One complete variation is called *one cycle*. The number of cycles per second is called the *frequency*. One cycle per second is called *one hertz*, abbreviated to Hz. The time taken for one cycle is called the *periodic time* and a moment's thought will indicate that

$$\text{periodic time (seconds)} = \frac{1}{\text{frequency}} \text{ (hertz)}$$

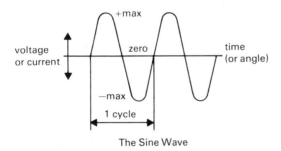

The Sine Wave

figure 4.11

The usual frequency of voltages generated commercially in the United Kingdom is 50 Hz, giving a periodic time of 1/50 second or 20 milliseconds. In the United States and Canada the frequency is 60 Hz, which is one reason why certain items of electrical equipment will not function equally well on both sides of the Atlantic.

Self-inductance and Mutual Inductance

A voltage is induced across a conductor whenever it is situated in a
magnetic field which is changing. In the generator, conductors are deliberately placed in a changing field in order to generate voltage; but the
effect of voltage induction is not confined only to generators. Since an
electric current sets up its own magnetic field surrounding the conductor
in which the current flows, whenever a current changes there will be an
induced voltage. Further, the induced voltage will act so as to oppose
whatever is causing its induction. The effect is called self-inductance
when the induced voltage is set up due to a conductor's own field.

All conductors have self-inductance. How large the effect is depends
on a number of variables: the rate of change of current and thus of flux
causing the induction, the strength of the flux caused by the current (as
determined by whether or not the conductor is wound in coil-form) and
also by the medium surrounding the conductor. If we wish the self-
inductance to be large, as we do on occasions, the conductor should be
wound in the form of a coil and the coil placed on a ferromagnetic core.
This increases the flux produced by the current so that when the current
changes the flux will be changing from a larger value and thus the rate of
change of flux will also be larger.

When considering self-inductance, since induced voltage depends on
rate of change of magnetic flux and magnetic flux is set up by electric
current then

  induced voltage depends upon rate of change of current

or more precisely

  induced voltage is directly proportional to rate of change of current

and writing this as an equation

  induced voltage = a constant × rate of change of current

The constant is called the coefficient of self-inductance of the conductor
and has the symbol L. It is measured in *henrys* (note the plural, the
singular being a henry, unit symbol H), where one henry is the coefficient
of self-inductance of a conductor when a current changing at the rate of
one ampere per second causes a voltage of one volt to be induced across
the conductor ends. The value of the coefficient of self-inductance
depends on the factors discussed above, namely, how the conductor is
wound, on what it is wound (the core material) and one other factor: the
state of magnetisation of the core. If a ferromagnetic material is used and
it is near saturation, so that flux cannot increase, this will affect the value
of flux and thus its rate of change.

The effect of self-inductance is noticeable particularly when circuits associated with magnetic fields are opened and closed. Equipment such as motors and relays require the setting up of a magnetic field for them to function correctly. Accordingly, whenever the circuit of the motor (or the field if this is being set up electromagnetically) or relay is opened, a voltage is induced which tries to maintain the state of affairs prior to the opening, that is, the back e.m.f. will try to keep current flowing. This effect is observable as sparks appearing across switch contacts.

Example 4.5

The self-inductance of a coil is 0.5 H. Find the back e.m.f. induced when a current of 2 A is reversed in the coil in 20 ms. Assume a constant rate of change of current with time.

Solution

Back e.m.f. = coefficient of self-inductance × rate of change of current

Since the current changes from +2 A to −2 A, the effective change is 4 A in 20 ms. The average rate of change is therefore 4 ÷ (20 × 10^{-3}) which equals 200 A/s, and

$$\text{back e.m.f.} = 0.5 \times 200$$

$$= 100 \text{ V}$$

The value here is fairly high for an induced voltage; much higher values can be obtained, of the order of kilovolts, for use in, for example, internal-combustion engine spark plugs.

If two conductors are placed so that the magnetic field set up by the current in one links the other conductor, then, when the current is changed, an induced voltage will be set up by the current across both conductors. The effect of a changing current setting up an induced voltage across an adjacent conductor is called *mutual inductance*. The induced voltage across the second conductor is related to the changing current in the first conductor by the equation

induced voltage across second conductor =
M × rate of change of current in first conductor

where *M* is the coefficient of mutual inductance between the conductors. If a current changing at the rate of 1 A/s in the first conductor induces a voltage of 1 V across the second, then the coefficient of mutual inductance between the conductors is 1 henry, the same unit as for the coefficient of self-inductance.

Example 4.6

The mutual inductance between two coils is 0.3 H. Find the e.m.f. induced across one coil when the current in the other changes uniformly from 6 A to 3 A in 10 ms.

Solution

$$\text{Change in current} = 3\text{ A in 10 ms}$$

$$\text{Rate of change of current} = \frac{3}{10 \times 10^{-3}}$$

$$= 300\text{ A/s}$$

$$\text{e.m.f. induced} = \text{coefficient of mutual inductance} \\ \times \text{rate of change of current}$$

$$= 0.3 \times 300$$

$$= 90\text{ V}$$

The coefficient of mutual inductance between coils depends on the medium in which the coils lie, which in turn determines the flux for a particular value of current. The mutual-induction effect is used in transformers where it is possible, using closely placed coils wrapped on a magnetic medium, to change voltage levels. The voltage to be changed, called the *primary* voltage, is connected to one coil producing a current and thus an interconnecting flux. The changing flux around the second coil induces a voltage called the *secondary* voltage. It can be shown that the ratio of primary voltage to secondary voltage is equal to the ratio of the number of turns on the primary coil to the number of turns on the secondary coil.

A Way of Defining the Weber

Since induced voltage across a coil is connected to the rate of change of flux around the coil by the equation

induced e.m.f = number of turns × rate of change of flux

if the voltage across a one-turn coil is 1 V and this voltage is induced by a changing flux, then the flux must be changing at the rate of 1 weber per second. One weber is thus one volt-second.

Transformers

As was stated earlier, when a magnetic field changes around a conductor a voltage is induced across it. If (as shown in figure 4.12) two coils of N_p

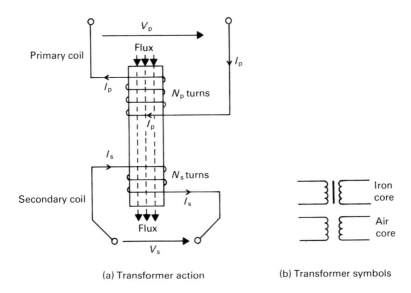

(a) Transformer action   (b) Transformer symbols

figure 4.12

and N_s turns respectively are linked by the same magnetic field set up by a voltage V_p applied to the coil having N_p turns, a voltage V_s is induced across the coil having N_s turns and the induced and applied voltages are related by the equation $V_s/V_p = N_s/N_p$; which can also be expressed as $V_s = V_p(N_s/N_p)$. Thus if a turns ratio (N_s/N_p) of 10/1 is used, 10 V may be induced in the secondary for every 1 V applied to the primary. (The coil across which the applied voltage is connected is called the primary coil; the other, the secondary coil.) The secondary current I_s is related to the primary current I_p by the equation $I_s/I_p = N_p/N_s$; which can be re-written as $I_s = I_p(N_p/N_s)$. For a 10/1 secondary/primary turns ratio, the secondary current would be 1/10 of the primary current. *Electrical power* is the product of voltage and current. It can be seen that V_s multiplied by I_s is $V_p(N_s/N_p)$ multiplied by $I_p(N_p/N_s)$, which cancels down to V_pI_p; from which it is clear that the output power (V_sI_s) is equal to the input power (V_pI_p).

Transformers are widely used in electronic systems for changing voltage and current levels and for *matching* one load to another.

Transformers for use at power and audio frequencies (to about 20 kHz) use metal cores for the magnetic circuit. At frequencies above this range dust cores and even air cores may be used because of the increasing losses at high frequencies.

Metal cores are made of high permeability steel alloys formed into thin laminations (about 0.35 mm thick) which are bolted together. Each lamination is insulated from its neighbour. The purpose of a laminated rather than solid core is to reduce the effect of all small circulating currents, called *eddy currents*, which are induced within the core by the changing flux. The core shape may be either of two shown in figure 4.13. Part (a) is generally known as 'core construction', and part (b) is known as 'shell construction'. The majority of power transformer cores use the configuration of figure 4.13a.

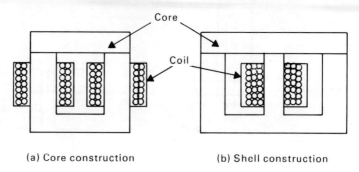

(a) Core construction   (b) Shell construction

figure 4.13

Transformer windings are generally circular in cross-section to withstand the considerable mechanical stresses set up when on load. There are various types of winding depending upon the intended use of the transformer.

Once assembled, the windings and core are contained within a metal case or tank. Cooling is effected either by air or, on larger power transformers, by oil, and provision is made for the coolant to circulate through the container.

5   Electrostatics and capacitance

Electrostatics is that branch of electricity concerned with static electric charge. We have seen that it is possible for materials to become electrically charged by gaining or losing electrons, one example being when certain materials are rubbed (glass with silk or ebonite with fur). It is also possible to charge materials using in the first instance moving electric charge, that is electric current, the charge then remaining on the materials.

If two pieces of conductive material are placed close to one another, separated by a thin layer of insulator, and the pieces are then connected to a source of electricity, such as a battery, electric current flows for a short period, the conductive pieces becoming charged in the process.

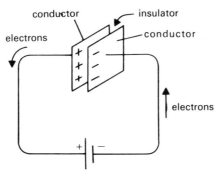

figure 5.1   Simple capacitor

Figure 5.1 shows such an arrangement. Electrons move from the negative pole of the battery to the right-hand conductor which becomes negatively charged. Similarly charged bodies repel one another (in much the same way as similarly acting magnetic fields) so the extra electrons on the charged conductor exert a repulsive force across the thin layer of insulator causing electrons to move off the left-hand conductor, leaving it positively charged. These electrons are then attracted to the positive pole of the battery so that electric current flows in the circuit connections. Note that electric current does *not* flow in the insulator between the conductors. Eventually the negatively charged conductor becomes so negative that further electron flow onto it stops because of repulsion between electrons. The two pieces of conductive material are now fully charged and a voltage is set up between them. If the battery is removed the arrangement continues to hold its charge for a long time, the time depending on how long it takes for discharge to take place through the surrounding air or other medium in which the materials are placed.

This arrangement is called a *capacitor* and the conductive materials, called *plates*, are said to have *capacitance*. The insulator between the plates is called the capacitor *dielectric*.

The amount of charge held by a capacitor is determined by the voltage across it and the capacitance of the capacitor. Capacitance is the ability to hold or store electric charge and is measured in coulombs of charge per volt applied. One coulomb per volt is called *one farad*, symbol F, the usual sub-multiples also being used. These are microfarad (μF) and picofarad (pF). The farad itself is relatively a large unit and capacitors normally have a capacitance ranging from a few picofarads to one thousand microfarads.

Denoting charge (coulombs) by Q, capacitance (farads) by C and voltage (volts) by V the equation relating these quantities is

$$Q = CV$$

Example 5.1

Calculate the charge held by a 10 μF capacitor when the voltage across it is 100 V.

Since   charge = capacitance × voltage

charge = $10 \times 10^{-6} \times 100$

= 10^{-3} coulombs or 1 millicoulomb

Example 5.2

Determine the voltage across a 1000 μF capacitor holding a charge of 100 mC.

Since $Q = CV$ (using the above symbols)

$$V = Q/C$$

So in this case $V = \dfrac{100}{1000} \times \dfrac{10^6}{1000}$

= 100 V

A capacitor offers an extremely high resistance to direct current since no current passes through the dielectric. The opposition to alternating current, called *capacitive reactance*, may be high or low and depends not only on capacitance but also on the frequency of the current. The effect of components in a.c. circuits are considered in detail later in the course.

Types of capacitor and symbols are described in chapter 11.

Capacitors are used for a variety of reasons including blocking d.c. (because of the high resistance), allowing a frequency-controlled passage of a.c. signals, storage and, because of the fact that a capacitor takes time to charge and discharge, the changing of the shape of voltage waveforms. Some evidence of this is shown later in the chapter when the effect of applying a direct voltage to a circuit consisting of capacitors and resistors is examined.

Series and Parallel Connection of Capacitors

The circuit illustrated in figure 5.2a shows a series connection of three capacitors C_1, C_2 and C_3 having voltages, V_1, V_2 and V_3 and a charge on each of Q coulombs.

Total voltage $V_{tot} = V_1 + V_2 + V_3$

$$= \frac{Q}{C_1} + \frac{Q}{C_2} + \frac{Q}{C_3}$$

(since voltage = charge/capacitance)

If

$$V_{tot} = \frac{Q}{C_{tot}}$$

where C_{tot} is the equivalent total capacitance then

$$\frac{Q}{C_{tot}} = \frac{Q}{C_1} + \frac{Q}{C_2} + \frac{Q}{C_3}$$

and

$$\frac{1}{C_{tot}} = \frac{1}{C_1} + \frac{1}{C_2} + \frac{1}{C_3}$$

or, in words, the reciprocal of the total equivalent capacitance of a circuit containing a number of capacitors in series is the sum of the reciprocals of the individual capacitances. Notice the similarity of the result for a series capacitor circuit to that for a *parallel* resistor circuit.

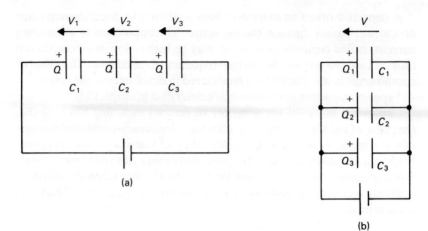

figure 5.2   (a) Capacitors in series; (b) capacitors in parallel

As might be expected the result for a parallel-capacitor circuit turns out to be similar to that for a series-resistor circuit.

For the parallel circuit in figure 5.2b the capacitors C_1, C_2 and C_3 are connected in parallel. The p.d. V is common and there is a charge on each capacitor, determined by the p.d. and the capacitance, of Q_1, Q_2 and Q_3.

Total charge

$$Q = Q_1 + Q_2 + Q_3$$
$$= C_1 V + C_2 V + C_3 V$$

and if $Q = C_{tot} V$, where C_{tot} is the total capacitance then

$$C_{tot} V = C_1 V + C_2 V + C_3 V$$

and

$$C_{tot} = C_1 + C_2 + C_3$$

or, in words, the total equivalent capacitance of a circuit consisting of a number of capacitors in parallel is the sum of the individual capacitances.

Capacitance and Resistance in Series d.c. Circuits

Figure 5.3a shows a series connection of a capacitor of capacitance C farads and a resistor of resistance R ohms connected via a switch to a d.c. supply of E volts. The voltage across a capacitor cannot change instantaneously since the capacitor must have time to charge, and charge and voltage are related to one another. The charge on a fully charged

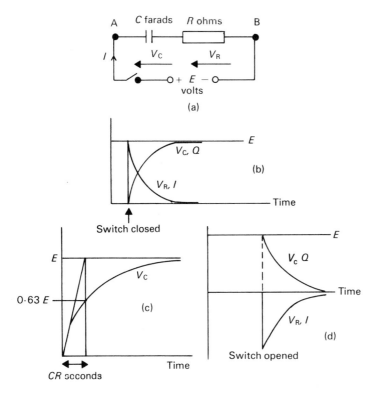

figure 5.3   CR circuits

capacitor depends upon the applied voltage and the capacitance. The time taken for this charge to accumulate must depend upon the total charge and thus upon the capacitance of the capacitor. Secondly, how long it takes for the capacitor to acquire the total charge also depends upon the charging current, the greater the current the smaller the time required. It will be recalled that electric current is in fact a measure of charge per unit time.

When the switch is closed there can be no charge on the capacitor at the instant of switching, thus the total voltage E must appear across the resistor R and the charging current at the instant of switching must therefore rise to E/R. This is the maximum possible value of current since once the capacitor begins to charge, its p.d. V_C will rise and V_R, which determines the current, must fall since at all times

$$E = V_R + V_C$$

With the initial surge of charging current the capacitor begins to charge quickly, but as carriers begin to accumulate on the plates the charging process slows down because of the repulsion between carriers and charge which must be overcome by the current.

A graph of charge Q, voltage V_C, voltage V_R and current I plotted against time is shown in figure 5.3b.

Notice that graphs of V_R and I have the same shape and graphs of Q and V_C have the same shape. Notice also that at all times $E = V_C + V_R$. The rate of charge is reflected in the slope of the charge/time curve and as can be seen it falls off with time.

This type of curve is well known in mathematics. It is called an *exponential* curve and the equation describing it contains the constant 'e', a constant which, like π, appears again and again in electrical engineering analyses. Tables of the value of e raised to various powers and logarithms of numbers to the base e are usually given in books of standard tables; e itself has the value 2.7183.

The time $t = CR$ is rather important.

If a tangent is drawn on the V_C/time curve at the point $t = 0$, it cuts the $V_C = E$ line at a time $t = CR$ as shown in figure 5.3c. This shows that CR is the time in which the capacitor would be fully charged if its initial rate of charge remained constant. In fact, it does not and at a time $t = CR$, V_C has reached about 63% of its final value of E volts, as shown.

The time CR seconds is called the *time constant* of the circuit. To use the equation time constant = CR seconds correctly, C must be in farads and R in ohms.

Theoretically V_C never quite reaches E but in practice it is assumed to do so after a time equal to $5 \times$ the time constant.

When the switch is opened, the capacitor theoretically will remain charged for ever. In practice it will discharge through the air, its rate of discharge being determined by atmospheric conditions. It is wise when examining circuits containing capacitors to ensure that when the supply is removed, the capacitors are fully discharged before touching. They may be discharged by placing a short circuit across their connecting leads.

For the circuit shown, if the switch is opened and a connection made between points A and B, the capacitor will discharge through the resistor. The exponential curves of the charging process are repeated; only the voltage V_C and charge Q are now decaying exponentially. The voltage V_R and the current I similarly fall to zero, but since the initial voltage V_R and current I are in the opposite sense to that during charging, the curves are as shown in figure 5.3d and the variable quantities appear to rise exponentially from a negative value to zero.

The time CR is still called the time constant and it is now the time for V_C to fall by about 63% of its original value.

The delaying property of capacitors in d.c. circuits, that is the way in which an instantaneous rise or fall in voltage can be prevented, is widely used in electronics.

Example 5.3

Calculate the effective total capacitance of a 0.1 μF capacitor and a 0.2 μF capacitor connected
  (a) in series
  (b) in parallel

(a) Series connection

let C_{tot} be the total capacitance; then

$$\frac{1}{C_{tot}} = \frac{1}{0.1} + \frac{1}{0.2} \quad \text{(where } C_{tot} \text{ is in microfarads)}$$

$$= \frac{2+1}{0.2}$$

$$= \frac{3}{0.2}$$

and $$C_{tot} = \frac{0.2}{3} \mu F$$

$$= 0.067 \mu F$$

Note in using the equation since the circuit capacitances were in microfarads, the total capacitance is also in the same unit. Care must be taken if the circuit capacitances are not in the same unit and before using the equation it is wise to write down all the values in farads (or in microfarads or picofarads), the result then being in whichever unit was chosen.

(b) Parallel connection

Using C_{tot} as before

$$C_{tot} = 0.1 + 0.2$$

$$= 0.3 \mu F$$

The same comment concerning units applies as in part (a).

Example 5.4

Calculate the resultant capacitance of a 1 μF capacitor connected in series with a 100 pF capacitor.

Now $$100 \, pF = 100 \times 10^{-12} \, F$$

$$= 0.0001 \times 10^{-6} \, F$$

$$= 0.0001 \, \mu F$$

Using C_{tot} as before:

$$\frac{1}{C_{tot}} = \frac{1}{1} + \frac{1}{0.0001}$$

(C_{tot} being in microfarads)

Thus
$$\frac{1}{C_{tot}} = \frac{1}{1} + \frac{10000}{1}$$

$$= 10001$$

and
$$C_{tot} = \frac{1}{10001}$$

$$= 0.0001 \; \mu F$$

and we see that the effect of connecting a relatively large capacitance in series with a small capacitance is negligible, the total capacitance being approximately that of the smaller of the two. This is a similar state of affairs to, say, connecting a 1 Ω resistor in *parallel* with a 1 MΩ resistor. The effective resistance of the parallel combination remains at approximately 1 Ω.

Example 5.5

Calculate the time constant of a circuit made up of a 0.5 μF capacitor in series with a 10 kΩ resistor.

$$\text{Time constant} = \text{capacitance} \times \text{resistance}$$
$$\text{(seconds)} \qquad \text{(farads)} \qquad \text{(ohms)}$$

Thus,   time constant $= 0.5 \times 10^{-6} \times 10 \times 10^3$

$$= 5 \times 10^{-3} \, s$$

$$= 5 \, ms$$

Example 5.6

A circuit made up of a 100 μF capacitor in series with a 1 MΩ resistor is connected across a 200 V supply. After what time will the capacitor voltage be approximately
    (a) 126 V
    (b) 200 V

(a) 126 V is 63% of 200 V and the time taken is equal to the circuit time constant

that is $100 \times 10^{-6} \times 1 \times 10^6$ s

This equals 100 s.

(b) The capacitor voltage may be assumed to reach the applied voltage after a time equal to 5 × time constant, that is 500 s.

And we see that this circuit with a large time constant takes over 8 min to reach a steady state.

6 Heat, light and sound

Heat

Part of the definition of energy is that it is the ability to change, convert or modify a body's state, shape or mass or state of rest. Heat is one form of energy and as such is measured in energy units: joules, the symbol for which is J.

Heat energy can be transferred from place to place by one or more of various methods considered in more detail below, and may be converted to other energy forms, including electrical, mechanical and chemical. Heat energy may be derived from other energy forms, as in the electric fire, the incandescent lamp, friction, impact of moving bodies, chemical reactions, and so on.

Heat Transfer

Heat energy may be transferred from one place to another in any of three ways. These are *conduction*, which takes place mainly in solids and liquids, *convection*, which occurs mainly in gases and liquids, and *radiation*, which requires no physical medium but can occur in solids, liquids and gases.

Conduction

All matter may be regarded as being made up of millions of tiny particles (molecules) which are constantly moving. These particles thus have energy of movement: kinetic energy. If a bar of suitable material is placed with one end in a heat source, such as a fire or flame, it is found that, after a time, the end remote from the heat source gets warm. The energy of the particles of matter at the heat source end is increased and some of this energy is passed to adjacent particles, until eventually the energy content as a whole rises, that is, the whole body gets warmer. No particles actually move down the bar although we sometimes use the expression 'heat flow' to describe the transfer of energy from one part to the other. This method of heat transfer is called *conduction*. It occurs mainly in matter which is made up of particles in close proximity, such as solids and liquids.

Convection

When heat energy transfer takes place because of actual movement of the particles containing the heat energy the process is known as *convection*. The best-known example is that of warm air, which moves from one place to another (usually rising) taking the heat energy with it. A movement of heated particles in the manner described is called a convection current; convection currents may be observed when heating a liquid by adding a coloured dye to the liquid.

Radiation

Heat energy may be *radiated* by means of *electromagnetic waves*. Whenever electric and magnetic fields act together such that one is at right angles to the other, energy is transferred. The nature of the energy depends on the characteristics of the electric and magnetic fields and may be heat, light, radio, ultraviolet, infrared, and so on. The energy we receive from the Sun, which is mainly light and heat energy, is conveyed in this manner, called *radiation*.

Radiation requires no physical medium but can pass through a number of different materials, depending on their atomic construction. Certain surfaces more easily absorb radiated energy while others reflect it. Black surfaces absorb and, if they are on a hot body, radiate more effectively than others; white or silvered surfaces reflect radiated energy. The reflector on an electric fire is silvered so that the radiated heat energy from the fire bar is emitted from the fire front. (It is interesting to note that so-called radiators in central-heating systems do not, in fact, radiate but transfer their heat energy by conduction and convection through the surrounding air.)

Temperature

Temperature and heat are not the same thing, although the terms are often loosely used to convey the same idea. The heat of a body is a measure of its total energy, whereas temperature ('the degree of hotness') is a measure of the *average* energy of the particles making up the body.

There are several *scales* of temperature available. A scale of temperature is a means by which temperatures may be compared with each other. Probably the best-known temperature scales are the Fahrenheit scale and the Centigrade scale, the latter being more correctly called the Celsius scale after the man who originated it. In devising the early temperature scales, much use was made of various temperature points thought to be fixed, the main ones being the boiling point of water and the melting point of ice. In fact, the temperatures at which these events occur depend on external physical conditions (the pressure of the surrounding air, for example) and are not suitable points for accurate definition. The Fahrenheit scale and the Celsius scale both took the melting point of ice and the boiling point of water as their fixed points, the range between the two being divided into a number of equal intervals called 'degrees'. The Fahrenheit scale has 180 degrees between the two points, the melting point of ice being taken as 32 degrees, written 32 °F, and the boiling point of water being taken as 212 °F. the Celsius scale puts the lower temperature point as zero, written 0 °C, and the upper as 100 °C. Both scales are still in common use, particularly the Celsius scale (although it is usually referred to as Centigrade) since, in an extended form, it is the SI temperature scale.

The temperature scale in SI units is the *Kelvin* scale. The fixed points on this scale are not those of the other scales but are much more accurately reproducible. They are *absolute zero* and the *triple point of water*. Absolute zero is the point at which an ideal gas is at zero pressure (an ideal gas being one in which the molecules are so small that their size can be ignored). The triple point of water is the point at which water, water vapour and ice exist in stable equilibrium. Between these points the temperature range is divided up into 273.16 equal intervals called *kelvins* (not degrees kelvin); absolute zero occurs at zero kelvin, written 0 K, and the triple point occurring at 273.16 K. At normal temperature and pressure the melting point of ice occurs at 273.15 K and the boiling point of water occurs at 372.15 K, so there are 100 kelvins between the melting point of ice and the boiling point of water. Clearly, then, one kelvin temperature interval is equal to one degree in the Celsius scale, so the Celsius scale can be used instead of the Kelvin scale, the difference being that temperature in Celsius is equal to temperature in kelvin *less* 273.15. Absolute zero occurs at −273.15 °C and the triple point at 0.01 °C.

Expansion Due to Heat

Many materials, particularly metals, change their shape when heated. In general this is due to each side of the body expanding as the heat energy is absorbed by the body material. If allowed to cool, the body resumes its original shape, provided that the increase in temperature has not been so great as to cause permanent distortion. One of the ways in which this fact is used in engineering is in *shrink fits*, where a piece of material to be joined to another is first heated, fitted over the other then allowed to cool. During cooling the piece shrinks and a tight fit is obtained. One example of the method is fitting steel 'tyres' to locomotive wheels.

Although only relatively small expansions are normally encountered (depending of course on temperature rise) they must be taken into account, and so measurements are made of expansions of different materials per temperature interval change. The expansion is taken into account in, for example, the building of bridges, buildings, railway lines and in machinery in general engineering.

Effect of Heat on Liquid and Gases

Solids have a defined shape, and in considering the effect of heat on them we consider linear expansion and expansion of area and volume. A liquid will change its shape and flow until it meets the side walls of the container in which it is situated. Consequently here we are not concerned with expansion of area or with linear expansion but only with volume change. Most liquids expand with increasing temperature at a constant rate over a particular range. One exception, however, is water which between 0 °C and 4 °C shrinks with increasing temperature and becomes more dense. This is why ice forms on the top of water, since the water at the slightly higher temperature than zero sinks and the colder water remains at the top. Gases behave differently to both liquids and solids, and expand to meet all container walls whatever the temperature, thus changing their volume and pressure as they do so. Pressure, volume and temperature of a gas are related by the general gas law.

Light

The sensation which we recognise as light is caused by electromagnetic waves with a frequency between 375 and 750 tHz (one terahertz is a million million cycles per second) and a wavelength between 760 and 380 nm (a nanometre is one-thousandth of a millionth of a metre). Radiation lying outside these limits have no visual effect on the human eye; those with wavelengths just longer give the sensation of heat and are called infrared, and those with wavelengths just shorter cause tanning of skin and, if intensive, can cause damage to body tissue; these latter rays are called ultraviolet. The light spectrum itself is divided into seven regions, each region having a different effect on the eye in terms of what we call 'colour'. These regions are red, orange, yellow, green, blue, indigo and violet. Red light is at the higher-frequency, longer-wavelength end of the spectrum and violet light is at the lower-frequency, shorter-wavelength end. White light is made up of all these radiations and, by a suitable arrangement of glass prisms and lenses, it can be split into its individual components. Electromagnetic waves in general are considered in chapter 7; in this chapter we shall take a closer look at the control of light using prisms and lenses and also mirrors.

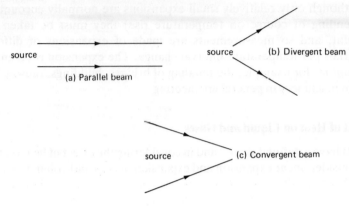

Light Beams indicated by Rays

figure 6.1

In determining the effect of materials on light waves it is useful to use the idea of a 'ray'. An assumption is made that light travels in straight lines (a close approximation) and lines are drawn from the light source indicating the direction of travel of the wave. These lines are called rays and the outer rays of a beam contain the beam as a whole, as shown in figure 6.1. The figure shows a parallel beam, where the outer rays never meet or move further apart, a convergent beam, where the outer rays eventually meet at some distance from the source, and a divergent beam, where the outer rays move further apart as they travel from the source. Using the processes of *reflection* and *refraction* of light rays, *images* of any object may be produced. An image in this sense is a visual impression of the object. Images produced by light rays may be larger or smaller than the original object and the study of the production of such images is called *geometrical optics*. Geometrical optics is used in engineering in the design, development and production of optical systems for use in cinema and television equipment, indicating instruments and illumination schemes, to name but a few.

Reflection and Refraction

When a light ray is *reflected* from a surface of a material it does not pass through the surface but returns from it either in the same or in a different direction to that in which it approached the surface. When a light ray is *refracted* by a material it passes through the material and its direction of travel is changed by the material. The change in direction is caused by the material affecting the speed of travel of the light waves.

Reflection at a Plane Surface

A plane surface is one which wholly contains every straight line joining any two points lying on it. Thus this page is a plane surface when it is lying flat but may not be so when it is turned, since a book page usually curves on turning. In diagrams any surface may be represented by the line which is seen when viewing one edge and thus a plane surface will then be represented by a straight line.

When light meets any surface some of it passes through and is refracted, the remainder is reflected. Some materials transmit more than they reflect, others reflect more than they transmit. As an example, a glass window transmits, but reflection does occur depending on conditions on each side of the window—a window at night behaves very much like a mirror, for example. A mirror, of course, reflects far more than its transmits.

Before taking a closer look at the theory of plane mirrors it should be noted that we see the colour of a material because of the white or almost white light falling on it; the material transmits most of the constituent light components but reflects the light corresponding to its colour. For instance, a red material will appear red when placed in white light because the remaining colours of the spectrum are transmitted or absorbed by the material whereas the red light is not. Illuminating any material with light which is not purely of one colour may cause the material to appear to change its colour, since the degree of transmission or absorption depends not only on the colour of the components of light but at what points in the spectrum the colours making up the light are placed. The effects are also determined to a large degree by individual colour sensitivity of the eye and brain making the observation. A white material reflects all the colour components of white light but the reflection is irregular and does not occur in the same way as it does with a mirror surface. A mirror surface produces a regular or *specular* reflection as described below.

Figure 6.2 shows a parallel beam of light entirely reflected by a plane surface. Since the beam is parallel, all the rays approaching the surface, which are called *incident* rays, are at the same angle to the surface. Reflection of each ray is symmetrical, so that the angle made by the reflected ray with the surface is the same as that made by the incident ray with the surface. Usually the angles we consider are not those made with the surface but are those made with a line perpendicular to the surface called the *normal*. As shown the angle of incidence and angle of reflection, both denoted by θ, are equal.

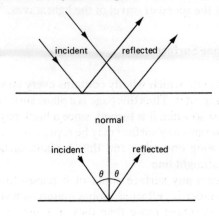

Reflection at a Plane Surface

figure 6.2

Figure 6.3 shows the formation of an image using a plane mirror. The object lies to the right of the mirror and two rays from any point on the object are drawn to the mirror surface. Each ray is reflected according to the laws of reflection described above. If now the reflected rays, which continue to diverge after reflection, are continued back behind the mirror, their continuations meet at a point as far behind the mirror surface as the object is in front. An eye observing the diverging rays, as shown, sees an image of the point lying behind the mirror and over all will see an image of the object as a whole.

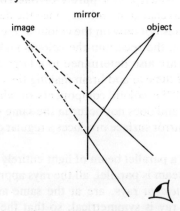

Image Formation in a Plane Mirror

figure 6.3

Since each point has an image behind the mirror at the same distance as that point is in front of the mirror, those points on the object lying furthest away from the mirror surface will produce images lying furthest away behind the mirror. This produces the effect known as *lateral inversion*, as shown in figure 6.4. It is most obvious if a mirror image of writing is produced when the mirror writing appears 'backwards'.

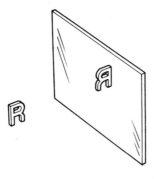

Lateral Inversion

figure 6.4

A mirror image produced by a plane surface is the same size as the object and is upright (or erect), that is, the 'same way up' as the object, and, because it is situated at a point from which light rays appear to come, it is called a *virtual* (as opposed to a *real*) image. A real image is one which could be cast on to a screen as by a film projector. Curved mirrors also produce images but they may be larger or smaller than the object, may be upright or inverted and may be real or virtual, depending on whether the mirror surface curves outwards towards the object (convex) or inwards away from the object (concave). A driving mirror is a convex mirror and gives a wide angle of view, although it is not as easy to judge distances with such a mirror as with a plane mirror, since the images are not necessarily the same distance from the mirror as the objects they represent. Curved mirrors are also used as reflectors in headlamps, spotlamps and torches. Here the curve is a particular type of curve called a *parabola* and the reflector is called a parabolic reflector. Parabolic reflectors produce a parallel beam of light from a source producing a divergent beam.

Refraction

In a vacuum the speed at which light and all electromagnetic waves travel—the velocity of propagation—is constant at 3×10^8 metres per second. When light enters a medium, however, it slows down and the amount by which the velocity of propagation is reduced depends on the nature of the medium. In some substances it becomes virtually zero and these substances can be used to block out light waves. If a parallel beam of light is incident normal to a surface, that is, enters at right angles to the surface, then it slows down as it passes through the medium, but retains its original direction both in the medium and when it emerges from the other side. If the same beam is not normal to the surface then the part of the wave front which arrives first begins to slow down before the remaining parts of the wave front which arrive later, and the overall effect is to make the beam deviate from its original direction. If the beam is travelling from one medium into another, and the second medium slows the beam, then the direction of travel moves towards the normal. If the beam enters from one medium into a second, which allows a higher velocity of propagation, the direction of travel moves away from the normal. The change in direction is called *refraction*.

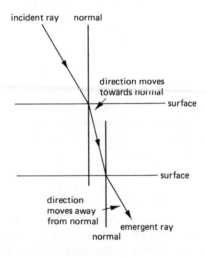

Refraction in a Parallel-sided Block

figure 6.5

Refraction of a ray of light as it passes through a block with parallel sides is shown in figure 6.5. The incident ray is not normal to the entry surface and, assuming a medium which slows the light, the direction moves towards the normal as the ray enters. On emergence into the original medium the ray direction changes again, moving away from the normal this time and, because the exit surface is parallel to the entry surface, the direction after leaving is the same as that before entering. If the sides are not parallel, the change in direction still occurs but the directions of travel before entering, while in, and after leaving the medium, are all different. Refraction of light is used to control direction, to focus beams and to produce images. To produce refraction we use a piece of optical apparatus called a *lens*.

Lenses

A lens is a piece of transparent material and usually has one or more curved surfaces. The most common shapes are the biconvex and biconcave, as shown in figure 6.6. The biconvex lens has two surfaces each curving outwards from the lens centre or *pole*. The biconcave lens has two curved surfaces curving inwards towards the pole. Other shapes, not illustrated, include the plano-convex, with one convex side and one plane side, the plano-concave, with one concave side and one plane side, the convex meniscus and the concave meniscus; the last two have two curved surfaces curved in the same direction, the difference being in the degree of curvature.

Types of Lens

figure 6.6

Effects of a Lens on a Parallel Beam

The effects of biconvex and biconcave lenses on an incident parallel light beam are shown in figure 6.7. The pole of each lens is denoted by the letter P. In the case of the biconvex lens the emergent beam is convergent and the light rays meet at point F, called the *focal point* of the lens. The emergent beam from a biconcave lens receiving a parallel beam is divergent and the focal point, F, in this case is the point from which the divergent beam appears to originate. It is obtained in a ray diagram by continuing back in the direction of the incident beam source the lines representing the emerging divergent rays. In both cases the distance between the pole and the focal point, PF in the figure is called the focal length. The focal length remains the same for these lens types regardless of the direction from which the incident beam approaches.

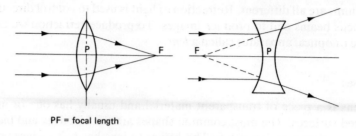

PF = focal length

Effect of Lens on Parallel Incident Rays

figure 6.7

Image Production by Lenses

If an object from which light is emitted, either directly or by reflection, is placed at an appropriate distance from a lens, an image is produced. As stated earlier, images may be real or virtual, erect (upright) or inverted and may be larger, smaller or the same size as the object. The nature of the image depends specifically on the type of lens and on the relative position of the object with respect to the focal point and the lens pole, that is, on whether the object lies within the focal length, outside the focal length or at the focal point. Figures 6.8, 6.9 and 6.10 show the production of images by various lens types with the object placed at different distances from the lens. In all these constructions the technique is to take a specific point on the object and draw two rays, one from the point through the lens pole and the other parallel to the lens principal axis (this being as shown in the diagrams), to the lens and from there to emerge in the new direction as determined by the lens type. The direction of the ray from the point on the object through the lens pole remains unaltered as indicated.

Figure 6.8 shows image formation by a biconvex lens when the object is situated at a distance along the principal axis from the pole greater than the focal length. In all such cases the image produced is real and inverted. Its size, however, depends on the actual amount by which the distance between object and lens exceeds the focal length. If the distance is greater than *twice* the focal length, the image is larger; if the distance is smaller than twice the focal length (but still larger than the focal length) the image is smaller. If the object placed at a distance along the principal axis exactly equal to twice the focal length the image is the same size as the object. Systems of this type producing smaller images are used in cameras, those producing larger images are used in enlargers or projectors and the type of system producing an image of the same size is used in an office copying-machine.

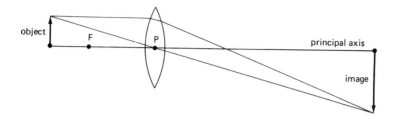

Image Formation by a Biconvex Lens: Object beyond Focal Point

figure 6.8

If the object is placed at the focal point the emergent rays are parallel and the image is said to appear at 'infinity', in other words, so far away that it cannot be seen. Such an arrangement is used in a spotlight to produce a parallel beam of light from a source emitting a divergent beam. The system can be compared with that using a parabolic reflector, in that both carry out the same function but, in the case of the lens arrangement, the emergent beam is on the opposite side of the optical element (the lens) to the incident beam from the object. In the mirror arrangement both emergent and incident beams are on the same side of the optical element, in this case, the mirror.

When an object is at a point on the principal axis between pole and focal point, that is, at a distance *less* than that of the focal length, the image produced is virtual, erect and magnified, as shown in figure 6.9. This arrangement is the principle of the magnifying glass.

A biconcave lens makes the rays which, when incident, are parallel to the principal axis, diverge on emerging and the image produced is always virtual, erect and reduced in size. The actual reduction depends on the exact position of the object relative to the lens pole. The situation is shown in figure 6.10 and the arrangement is used in spectacles and similar systems.

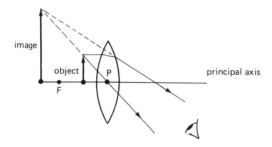

Image Formation by a Biconvex Lens: Object between Focal Point and Lens Pole

figure 6.9

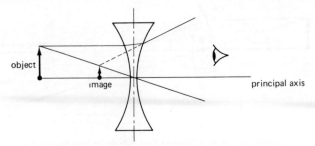

Image Formation by a Biconcave Lens

figure 6.10

Colour Mixing

Colour, like light which produces the effect, is a visual sensation. Light waves strike the eye and the optic nerve transmits signals to the brain producing the effect we call 'light' and also the effect we call 'colour'. White light is in fact made up of seven colours, red, orange, yellow, green, blue, indigo and violet and this can be seen by passing white light through a specially shaped piece of glass called a prism, as shown in figure 6.11. As was stated earlier light itself is an electromagnetic wave, similar to a radio wave, and is produced by electric and magnetic fields changing in strength at regular intervals. The frequency at which the fields change their strength determines the kind of wave and the effect it has. The range of frequencies of electromagnetic waves is called the electromagnetic spectrum and light waves occupy only a very small part of this, lying, in terms of frequency, just above heat waves (producing heat radiation) and just below ultra-violet rays (producing tanning of the skin). The band of light frequencies may be divided into seven, corresponding to the seven colours that make up white light, the low frequency end producing the colour red (hence the top end of heat radiations is called infrared, infra meaning 'below'), the top frequency end producing the colour violet (hence the radiations above this being called ultraviolet, ultra meaning 'above').

When white light is passed through a prism, each component wave is affected differently or, to be precise, is refracted differently. Hence at the other side of the prism the component waves are separated and produce the seven different colours. This effect is sometimes seen when rain and sun occur together, light from the sun being split by the millions of water drops in the air to give the familiar rainbow.

As white light is made up of seven colours so certain other colours can be made up by mixing two or more of the seven. Yellow light, for example, may be made up by mixing red light and green light. We are

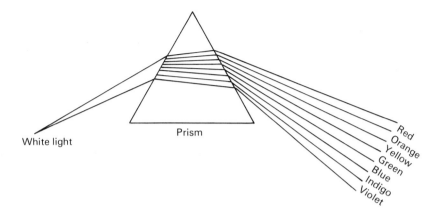

White light

Prism

Red
Orange
Yellow
Green
Blue
Indigo
Violet

figure 6.11

going to examine more closely what is meant by 'mixing' light but first we must look at why an object appears to have colour. Transparent objects, that is, those made of a material which allows light rays to pass through them, have colour because they allow through only the light rays making up that colour. A piece of red glass allows the passage of only red light, a piece of yellow glass allows the passage of only red light and green light, the effect on the eye in this case being that we call 'yellow'. Opaque objects, those made of a material which does not allow the through passage of light rays, appear to have colour because they reflect only those light rays making up the colour and absorb (but do not allow through passage of) all other light rays. Thus if white light is shone on a yellow object, all light rays except those producing red and green, are absorbed, the red and green rays being reflected back to the eye and giving the effect of yellow.

There are two ways in which light of different colours may be produced, *additive* mixing and *subtractive* mixing. Additive mixing is carried out by the addition of light rays of different colours to produce the resultant colour effect on the eye. Thus if red light is shone on to a screen and green light is shone on to the same part of the screen the resultant effect is yellow. In determining the resultant effect of mixing whether additive or subtractive it is useful to remember the following combinations

red, blue and green produce white
red and green produce yellow
blue and green produce cyan
red and blue produce magenta.

Red, blue and green are called the *primary* colours of the spectrum since each one cannot be produced by additive mixing of two or more of the others. White light can be produced not only by mixing the primary colours but by mixing one of the primary colours with one of the other colours containing the other two primaries. For example:

red and cyan produce white
(because cyan is produced by mixing blue and green)

and green and magenta or blue and yellow. These pairs are called complementary colours. Additive mixing of light of different colours is used in colour television and colour photography. One method used in colour television is to view with the camera the scene being televised through a red filter, a blue filter and a green filter in turn so that three electronic signals are produced corresponding to the red light, the blue light and the green light from the scene. These signals are then processed and used eventually to cause dots of red, blue and green light to appear on the television screen. The light from the dots is combined via the eye and brain to produce a picture of the same relative colours as the scene being televised.

When light filters are used or when pigments (paints) are mixed to produce a resultant colour the mixing process is called subtractive because the principle involved is that of removing from white light the colours which are not required (rather than adding colours to produce the resultant effect).

Thus if white light is shone on to a yellow filter we have red, blue and green light (which are the primary components of white) coming to the filter but only red and green light (the primary components of yellow) passing through. The blue light, to which a yellow filter is opaque, has been subtracted from the white light. If now a magenta filter is placed behind the yellow filter only red light is allowed through because magneta is made up of red and blue and a magenta filter is opaque to green. Thus the red and green light from the yellow filter enters the magenta filter but only red leaves it. Again the green component has been subtracted from the yellow. A similar subtractive mixing occurs when pigments are mixed to produce paints of a particular colour. For example, a yellow pigment is one which absorbs the blue component of white light and reflects the red and green to produce yellow, a magenta pigment (sometimes incorrectly referred to as red) is one which absorbs the green component of white light and reflects the red and blue to produce magenta. If these two pigments are mixed they absorb together both the blue and green components of white light and reflect red. It should be noted that the colours yellow, magenta and cyan, often erroneously called yellow, red and blue, are sometimes known as *artist's primary colours*, or just *primary colours*, because they are the basic colours used in the mixing of pigments. The true primary colours in the light spectrum, however, are red, blue and green, the word primary having the meaning given earlier, that is these colours cannot be produced by the mixing of light of other colours.

The basic techniques of a colour television system are discussed in chapter 10.

Sound Waves

Sound is a *pressure* or *compressional* wave. Vibration of the source, that is, an actual movement to and fro or up and down, alternately *compresses* and *rarefies* the matter next to the source, and this compression and rarefaction is passed along from particle to particle along the path of the sound (see figure 6.12). Hence a vacuum or a near-vacuum has insufficient particles of matter to transmit the wave and so sound cannot travel in such conditions. This fact is well illustrated by placing an electric bell inside an inverted bell jar and evacuating the jar. As the air becomes more and more rarefied the sound dies but we can see that the bell is still ringing by the movement of its clapper.

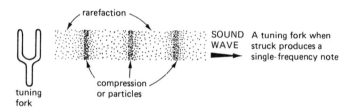

figure 6.12

Sound frequencies detected by humans range from about 20 Hz to 20 kHz, but both limits vary considerably from person to person, being dependent on age, health, and so on. Certain animals, noticeably dogs and bats, hear higher frequencies (hence 'dog whistles' inaudible to humans); frequencies immediately above 20 kHz but below normal radio frequencies are classified as *ultrasonic*. Ultrasonic waves are very useful in intruder alarms, because they are virtually undetectable by humans; they are also used in industrial cleaning processes.

Sound waves have a velocity dependent on medium; in air it is approximately 333 m/s depending on temperature. Supersonic velocities are often expressed as a ratio called *Mach number*, where

$$\text{Mach number} = \frac{\text{velocity of body}}{\text{velocity of sound in the same medium}}$$

'Mach 1' means that a body is travelling at the velocity of sound, 'Mach 2' that it is travelling at twice the velocity of sound, and so on.

When a sound wave is produced by vibration, the frequency of the sound is determined by the vibrating source characteristics, for example, the shape and size of, say, a 'tuning fork' (used to provide a single-frequency sound source), the tension and mass/length of a musical instrument string, and the length and diameter of pipes used in organs and other wind instruments.

7   Electronic signals

The purpose of an electronic system is to convey intelligence from one point to another, the word 'intelligence' meaning information and including words, music, pictures and numerical data. Intelligence is conveyed from source to receiver by means of *electronic signals*.

d.c. Signals

If direct current or voltage, that is, current flowing or voltage acting in one direction, is used to convey intelligence it is called a d.c. signal. Figure 7.1 shows a simple electrical circuit made up of a d.c. power supply, a d.c. electric motor and a switch. When the switch is closed direct current flows in the circuit and the motor runs. The intelligence conveyed is the instruction 'run', the signal being the direct current. When the switch is opened, the motor stops running, the intelligence conveyed being the instruction 'stop'. A graph plotting voltage against time is shown in the figure and the 'run' and 'stop' intervals are shown quite clearly.

If the simple switch in the circuit of figure 7.1 is replaced by a double-pole, double-throw switch, as shown in figure 7.2, it is now possible to reverse the direction of current flow and the intelligence can now consist of one of three instructions: 'run in one direction', 'run in the reverse direction' and 'stop'. The signal is still a d.c. signal since it is in one direction at the time of sending the instruction.

A more complex set of instructions may be conveyed to the motor if a variable power supply is now introduced, capable of providing different levels of voltage and current. Figure 7.3 shows such a system, the instruction set now including the commands 'run in one direction', 'run in the other direction' (each of these being at any one of a set of possible speeds), 'stop', 'go faster' and 'go slower' (each of these being in one of two possible directions). Motor control using d.c. signals is used in a wide variety of situations, a few being steel mills, traction, automatic door opening, conveyor belts, pen recorders and so on.

d.c. signals are also used in light exposure meters, in which a direct current affected by the light level sets the camera, in code transmission (for example, Morse as shown in figure 7.4) and in computers and other logic systems. The transmission of d.c. signals always requires wires between source and receiver and the changing of levels of voltages and currents used as signals is usually more difficult than with other forms of signal.

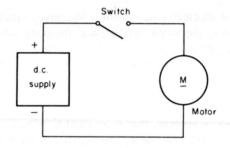

(a) Simple electrical circuit using single-pole, single-throw switch and d.c. motor.

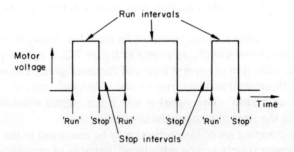

(b) Graph of voltage received by motor in circuit of figure 1.1 (a)

figure 7.1   The unidirectional d.c. signal

a.c. Signals

An alternating current or voltage is one which changes its direction at regular intervals. As we have seen alternating current and voltage levels can easily be changed using a transformer and, as will be explained shortly, if the rate of change is fast enough it is possible to convey intelligence using alternating currents and voltages without having connecting wires between source and receiver. When alternating currents and voltages are used to convey intelligence they are called a.c. signals.

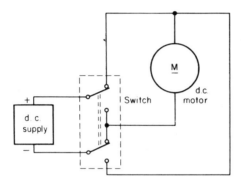

(a) Motor circuit with reversing facility

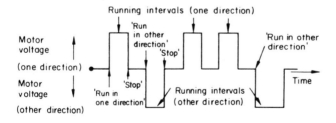

Motor is stopped when graph lies along horizontal axis

(b) Voltage versus time graph
for circuit of figure 1.2 (u)

figure 7.2   The two-directional d.c. signal

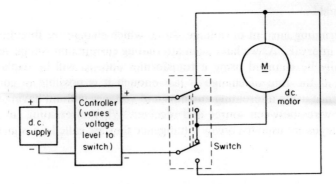

(a)   Motor circuit with speed adjustment and reversing facility

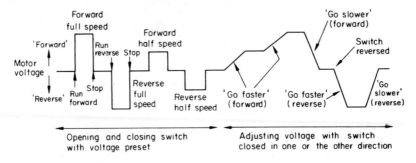

(b)   Graph of voltage against time for circuit of figure 1.3 (a)

figure 7.3   Two-directional d.c. signal with variable voltage

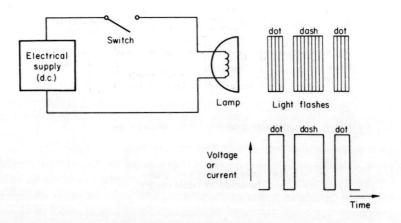

figure 7.4   Conveying verbal intelligence using Morse code

Earlier we looked at a d.c. circuit containing a motor, d.c. power supply and reversing switch and here the current flowed in one of two possible directions. This is not an a.c. signal, however, since when the current changes direction the instruction conveyed changes from 'run in one direction' to 'run in the reverse direction'. An alternating current or voltage changes at regular intervals but the intelligence conveyed does not necessarily change each time the direction of current flow (or of action of the alternating voltage) changes. A simple illustration of this is using an a.c. signal to produce a warning note in a paging or security system. A single tone in a loudspeaker is produced by a constant frequency signal, the voltages and currents reversing at regular intervals but the intelligence (the warning note) is the same. If the frequency is increased a higher note is heard, the signal changing at a higher rate but again a single unchanging (higher) note being the intelligence.

a.c. signals are widely used for intelligence transmission. The best known systems are probably domestic radio, television and sound reproduction but there are many other applications including the transmission of numerical and coded data between a central control and remote vehicles or stations. To examine more closely the uses of a.c. signals it is first necessary to look at different types, particularly electromagnetic waves and the electromagnetic spectrum.

The Electromagnetic Spectrum

a.c. signals using only current require conductive paths in exactly the same way as d.c. signals and source and receiver need to be connected by wires. However, it is found that if an alternating current changes fast enough it is possible to transfer energy, and thus intelligence, from one circuit to another which is not linked by wires to the first. This is due to the setting up of electromagnetic energy waves which radiate from one circuit and are received by the other.

To obtain a picture of energy transmission by wave motion consider the following examples: the first shows the creation of a wave and how it travels, the second shows how such a wave carries energy. Imagine a rope tied at one end to a wall, the other end being held in the hand. If the hand is now moved up and down, as shown in figure 7.5 a wave travels down the rope towards the fixed end. Each part of the rope remains in the same vertical plane, but the wave travels in the horizontal plane. The second example is that of a stone dropped into a pond. Waves radiate radially from the point of impact, the propagation taking place in much the same way as in the previous example. Any object floating near by will absorb energy and rise and fall. This example illustrates the transfer of energy from the stone to the floating object via the water wave.

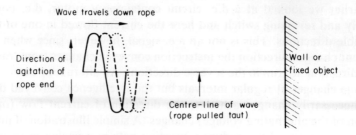

figure 7.5   Generating a rope wave

Whenever a current is alternating within a conductor an electric field is set up around the conductor. This field builds up in strength and then collapses as the current changes. In addition, a magnetic field is also set up, at right angles to the electric field and the magnetic field strength also increases and reduces alternately. The combination of the changing electric and magnetic fields sets up an electromagnetic wave which radiates from the conductor carrying the current in much the same way as the water wave discussed earlier. The distance covered by the radiation and the strength of the electromagnetic wave generated is determined (among other things) by the conductor length, by the rate of change and magnitude of the alternating current and by the associated voltage. The topic of propagation is dealt with more fully in chapter 8.

It should be noted that an electromagnetic wave is not exactly the same as those discussed earlier, for both the water wave and rope wave travel within a medium (that is water, or rope, respectively) whereas an electro-magnetic wave can travel within a vacuum, because the electromagnetic wave is a rise and fall in the strength of electric and magnetic fields and these do not require matter for their existence. However, the alternate strengthening and weakening of the fields does result in a wave pattern that travels outwards from the source and thus can be used for the transmission of energy and intelligence.

The wave pattern of an electromagnetic wave is similar to the one already given for an alternating current contained within a circuit (that is, a sine wave); but whereas the sine wave depicting an alternating current represents current magnitude plotted against time, the wave shape of an electromagnetic wave can be regarded as a plot of field strength (as determined by the combination of electric and magnetic fields that produced the wave) against distance travelled. The two waveforms in transmission are, of course, closely related, as it is the alternating current within the circuit which produces the electromagnetic wave outside the circuit.

There are several important characteristics of a wave, certain of which are common to both the a.c. sine wave and the electromagnetic sine wave and certain of which are more commonly used with one or other of the two waves as explained below. The important characteristics are amplitude, wavelength, periodic time and propagation velocity (see figure 7.6).

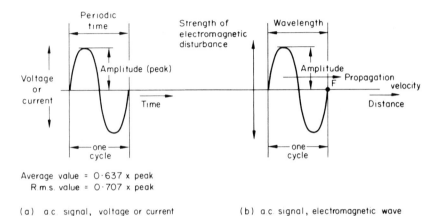

Average value = 0·637 x peak
R.m.s. value = 0·707 x peak

(a) a.c. signal, voltage or current   (b) a.c. signal, electromagnetic wave

figure 7.6   Types of a.c. signal wave

The *amplitude* is the peak value reached in either direction from the centre of the wave. When the sine wave represents an alternating current or voltage other values of magnitude are also important. These are the average value and the root-mean-square or r.m.s. value. The average value of a complete cycle is, of course, zero as there are equal excursions above and below the centre line. For a sine wave the average value of half a cycle is the one normally taken and it can be shown that it is in fact equal to 0.637 of the peak value of amplitude. The average value is the one that determines the deflection of certain instruments such as those of the moving coil variety which are used for current or voltage measurement. The r.m.s. value is the value of the equivalent direct current or voltage to do the same work (that is, provide the same energy) under the same circumstances as the alternating current or voltage. It can be shown that the r.m.s. value of a sine wave is equal to 0.707 of the peak value. The r.m.s. value is the one most commonly used when referring to alternating currents or voltages—the mains voltage in the United Kingdom, for example, being given as 240 V, has an r.m.s. value of 240 V and thus a peak value of 240/0.707, that is 339.4 V. The average value of the mains voltage over one cycle would then be 339.4 × 0.637 that is 216.2 V. The ratio between r.m.s. value and average value of any waveform is called the *form factor*. The form factor for a pure sine wave is thus 0.707/0.637, that is 1.11.

The *frequency* of a wave is the number of cycles per second, where a cycle is a complete alternation from zero to maximum value in one direction, back through zero to maximum value in the other direction and finally back to zero (see figure 7.6). One cycle per second is called one *hertz* (abbreviated Hz) and the multiple units include the kilohertz (kHz) which is 1000 cycles/second, the megahertz (MHz) which is 1 000 000 cycles/second and the gigahertz (GHz) which is one thousand million cycles/second. The frequency of the a.c. mains supply in the UK is 50 Hz and in the USA and Canada is 60 Hz. The frequency of a wave is extremely important as to a great extent it determines the use to which the wave can be put.

The *periodic* time is the time taken for one cycle. A moment's thought will show that the periodic time is the reciprocal of the frequency, for if the frequency is f cycles per second (hertz), then the time taken for one cycle must be $1/f$ seconds. If the wave is of an alternating current or voltage then the periodic time can be shown directly on the graph as in figure 7.6a. It cannot be shown directly on the graph of an electromagnetic wave, because distance and not time is plotted along the horizontal axis.

The *wavelength* is a term used particularly when discussing electromagnetic waves and is the distance occupied by one cycle (see figure 7.6b).

The *propagation velocity* is again a term used in connection with electromagnetic waves and is the velocity with which a wave travels. In fact, all electromagnetic waves travel at the velocity of light, namely 300 million metres per second (about 186 thousand miles per second). The wave front of an electromagnetic wave (that is the point F in figure 7.6b) travels a distance equal to one wavelength in the time taken for one cycle. In one second the number of cycles formed is known as the frequency. Thus the distance travelled by the wave front in one second is equal to the product of frequency and wavelength. The distance travelled in one second is, of course, the propagation velocity so that propagation velocity = frequency × wavelength; and since the propagation velocity is constant, frequency is proportional to 1/wavelength. For all electromagnetic waves we see therefore that as the frequency is increased the wavelength is reduced; that is, long waves are at low frequencies, short waves are at high frequencies.

The complete range of electromagnetic waves varies from very low to very high frequencies (very long to very short wavelengths) and is called the electromagnetic spectrum. Details of the spectrum are given in table 7.1. Broadly speaking, the range may be divided into audio waves, radio waves (including television and radar waves), heat waves, light waves and the upper region of frequency which yields ultraviolet rays, X-rays and cosmic rays. (The word wave is replaced by ray in this region.)

	Frequency (Hz)		Wavelength (m)	
	3.0×10^{22}		10^{-14}	
	3.0×10^{20}		10^{-12}	
	5×10^{19}		6×10^{-12}	Cosmic rays
Gamma rays	1.5×10^{18}		2×10^{-10}	
	2.5×10^{16}		1.2×10^{-8}	X-rays
Ultraviolet rays	3×10^{15}		10^{-7}	
	7.5×10^{14}		4×10^{-7}	
	3.75×10^{14}		8×10^{-7}	Light waves
Infrared (heat)	3×10^{12}		10^{-4}	
waves	7.5×10^{11}	Hertzian	4×10^{-4}	
Radar, TV, VHF	8.9×10^{8}	(radio)	3.37×10^{-1}	These are
Radio bands	4.7×10^{8}	waves	6.38×10^{-1}	sample fre-
	5.4×10^{7}		5.55	quencies within the band
Short wave	1.6×10^{6}		187.5	
Medium long waves	2×10^{4}		1.5×10^{4}	Limit of
	20 Hz		1.5×10^{7}	human ear

10^{12} means 1 000 000 000 000 (that is 1 followed by 12 zeros).
10^{-12} means 0.000 000 000 001, that is, zero followed by 11 $(12 - 1)$ zeros; in other words there are $(12 - 1)$ zeros following the decimal point.

table 7.1   The Electromagnetic Spectrum

Audio waves are those which when received and processed by suitable equipment give rise to sound waves that can be heard by the ear. Audio waves as such cannot be heard because they are solely an electromagnetic disturbance; however, if alternating currents changing at frequencies within the audio range are fed to a loudspeaker, sound waves are generated which can be heard. Sound waves as distinct from electromagnetic audio waves were considered in chapter 6.

Electromagnetic waves at audio frequencies are difficult to propagate and it is usual to employ higher frequency waves in the radio frequency band to carry audio intelligence, using a process called modulation. This process is described later in the chapter. The range of frequencies used as carriers, sometimes called *Hertzian* waves after their discoverer, extends from just above the audio range at about 20 kHz to just below visible light, the infrared region beginning at around 750 GHz. Above the radio wave region there are radiant heat waves, light waves and the ray region. The visible light spectrum is quite small and it should be noted that, unlike electromagnetic audio waves, *electromagnetic waves at light frequencies* do produce the sensation of light directly on the human sensory organ, the eye.

As the frequency is further increased above the light region the waves become more penetrating and can cause physical changes in the human body. Ultraviolet radiation causes skin changes which in the mild form result in tanning of the skin and in the severe form can result in burning. The ability of X-rays to penetrate human tissue is well known and intense forms of radiation at these frequencies are used to destroy living matter. The effects of prolonged exposure to cosmic rays, from which earthbound humans are protected by the atmosphere, are still being investigated during flights above and beyond the earth's atmosphere.

Non-sinusoidal Waveforms

Figure 7.7 shows the waveforms of alternating currents which, if supplied to a loudspeaker, would yield the sounds 'o' and 'oo'.

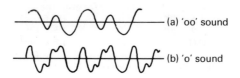

figure 7.7   a.c. waveforms corresponding to certain sounds

As the resultant sound is composed of a number of component vibrations, so the a.c. signal which yields the sound can be considered to be made up of a number of component a.c. signals, each of which is of pure sine-wave form. Figure 7.8 is a further example of this and shows how an approximate square wave is made up of a number of component sine waves. In this example the frequency of each of the component sine waves is a simple multiple of the frequency of the largest component. The frequency of the largest component is called the fundamental *frequency* and the components are called *harmonics* of the fundamental frequency. The harmonic which has a frequency equal to twice the fundamental

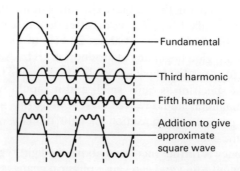

figure 7.8   Complex waveform breakdown in harmonic components

frequency is called the second harmonic, the harmonic which has a frequency equal to three times the fundamental frequency is called the third harmonic and so on. Figure 7.8 shows that the sum of the fundamental frequency and its odd harmonics (3 × fundamental frequency, 5 × fundamental frequency, 7 × fundamental frequency etc.) is a square wave. Note that the amplitude of each harmonic is progressively reduced as the order (that is, the number) of the harmonic is increased. It can be shown similarly that the sum of a fundamental frequency and its even harmonics yields a triangularly shaped wave.

The wave characteristics such as frequency, peak value, average value etc., that are used to define a particular sine wave are also used to define regularly occurring non-sinusoidal waves, although, of course, the relationship between these values is different, for example the average value of a half-cycle of a sine wave is 0.637 of its peak value but the average value of a half-cycle of a square wave is 0.5 of its peak value. Figure 7.9 shows square and triangular waves, each composed of a fundamental and a large number of harmonic sine waves (the larger the number the 'purer' is the resultant wave shape) and various characteristics are shown in the figure.

An important characteristic of a waveform consisting of a number of square or rectangular pulses especially when the waveform has different widths of pulse, in the *mark to space* ratio. This is the ratio between the pulse width and the width of the no pulse region.

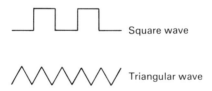

figure 7.9

If the total cycle time is made up of a pulse taking up two-thirds of the time and no pulse for the remainder, the mark to space ratio is 2 : 1 (see figure 7.10).

It is useful to remember that any complex waveform can be broken down into components. The shape of any signal wave can be changed as desired by passing the signal through suitable circuitry which affects one or more of the components more than others. Such circuits include *integrating, differentiating* and *filter* circuits. Also, it is possible to obtain a resultant complex waveform by summing components (not necessarily harmonics). This process takes place during *modulation*, which is the process used to impress intelligence upon a carrier wave. This is discussed below.

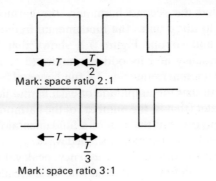

figure 7.10   Mark to space ratio

Modulation

As already stated, it is difficult to propagate electromagnetic waves at audio frequencies. This is largely due to the length of the aerial required and the high power that would be necessary to obtain efficient transmission. The normal method is to use electromagnetic waves at radio frequencies as carrier waves and to modulate (modify) these carriers with the intelligence signal to be transmitted. There are three kinds of modulation in common use: *amplitude modulation* (AM); *frequency modulation* (FM) and *pulse modulation*. In amplitude modulation the intelligence signal is used to change the amplitude of the carrier wave. The carrier amplitude then varies as the amplitude of the signal to be carried. This is shown in figure 7.11a. The rate of change of the carrier amplitude is determined by the frequency of the modulating signal; the frequency of the modulated signal is held constant. The percentage of modulation of an amplitude modulated carrier is the amount by which the carrier amplitude changes relative to the amplitude of the unmodulated wave, expressed as a percentage. This is shown in figure 7.11b which shows a carrier modulated at different percentages. As can be seen 50% modulation means that the carrier amplitude rises 50% above the unmodulated maximum value when the modulating signal rises to its maximum value in the 'positive' direction and the carrier amplitude falls below the unmodulated maximum by the same amount when the modulating signal increases to its maximum value in the 'negative' direction. The maximum percentage modulation which can occur without distortion is 100% when the carrier rises to twice the unmodulated maximum and falls to zero. Modulation above 100% distorts the carrier waveform from a pure sine wave and leads to the presence of undesirable harmonic frequencies.

Amplitude modulation gives a carrier of complex waveform and (as

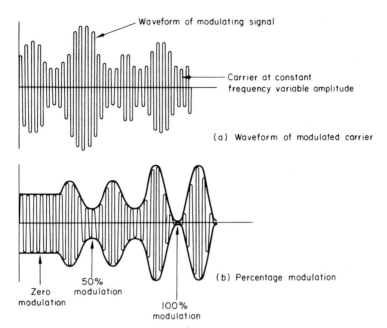

Waveform of modulating signal

Carrier at constant
frequency variable amplitude

(a) Waveform of modulated carrier

(b) Percentage modulation

Zero
modulation

50%
modulation

100%
modulation

figure 7.11   Amplitude modulation

explained above) such a waveform can be considered to be made up of
several component waves at different frequencies. It can be shown that if
an r.f. carrier of frequency f_c is amplitude modulated by a pure sine wave
of frequency f_m, the resultant waveform contains three components at
frequencies of $f_c - f_m$, f_c and $f_c + f_m$ respectively. For example, if a
1000 kHz carrier is amplitude modulated by a signal of frequency 4 kHz,
the resultant modulated wave would contain three components at frequencies 996 kHz, 1000 kHz and 1004 kHz. If the modulating signal
is itself complex (that is, it contains many frequencies), two side
frequencies are generated for each component frequency of the modulating signal. Thus, the resultant modulated wave will contain a number
of frequencies between the limits $f_c - f_{max}$ and $f_c + f_{max}$ where f_{max} is the
highest frequency present in the modulating signal. The band of
frequencies lying between the carrier frequency and the upper limit $f_c +
f_{max}$ is called the *upper sideband* and the band of frequencies between the
carrier frequency and the lower limit $f_c - f_{max}$ is called the *lower sideband*
(see figure 7.13). Circuits handling the modulated wave must be able to
handle all component frequencies in both sidebands if faithful reproduction of the intelligence is to be obtained at the receiver. The *bandwidth* of such circuits (that is the band of frequencies that can be handled)
must then be equal to $(f_c + f_{max}) - (f_c - f_{max})$, that is $2f_{max}$ (or twice the
highest modulating frequency).

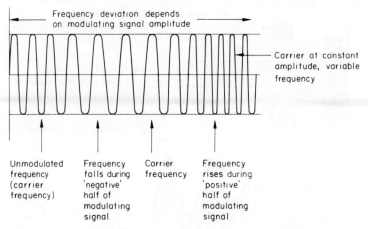

figure 7.12   Frequency modulation

A *frequency-modulated* carrier wave is shown in figure 7.12. In this type of modulation the carrier amplitude remains constant and the carrier frequency changes. The total change or *deviation* of the carrier frequency is determined by the amplitude of the modulating signal and the rate of change of the carrier frequency is determined by the frequency of the modulating signal. As the modulating signal waveform increases in the 'positive' direction, the carrier frequency increases; and as the modulating signal waveform increases in the 'negative' direction, the carrier frequency falls. The range of carrier frequency from centre (unmodulated frequency) to highest, or from centre to lowest, is called the *maximum frequency deviation*.

As the carrier frequency is constantly changing the waveform cannot be a pure sine wave once the carrier is frequency modulated. The modulated waveform is thus complex and contains a number of component or side frequencies. With amplitude modulation each frequency present in the modulating signal generates two frequencies in the modulated carrier, as already explained. With frequency modulation, however, each frequency in the modulating signal generates a large number of frequencies in the modulated carrier. Using the notation above, a carrier of frequency f_c modulated by a single frequency signal f_m would contain side frequencies at $f_c + f_m, f_c - f_m, f_c + 2f_m, f_c - 2f_m, f_c + 3f_m, f_c - 3f_m, f_c + 4f_m, f_c - 4f_m$ and so on. The actual number is determined by the extent of the carrier wave distortion and thus depends upon the maximum frequency deviation. As this, in turn, depends upon the amplitude of the modulating signal it is this factor which determines the number of side frequencies. The number generated for any given case can be calculated using a form of advanced mathematics known as Bessel functions, the theory of which is not within the scope of this book.

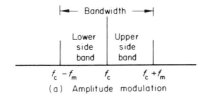

(a) Amplitude modulation

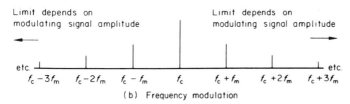

(b) Frequency modulation

figure 7.13   Side bands: vertical lines indicate relative amplitudes (not drawn to scale)

The highest or lowest side frequency generated per modulating frequency must not be confused with the highest or lowest frequency to which the carrier swings during modulation. The side frequencies are the frequencies of the pure sine waves which, when added together, give the complex waveform of the modulated carrier. The highest or lowest frequency to which the carrier swings is the frequency of alternation of the complex carrier waveform (a distorted sine wave) during a fractionally small interval of time. The side frequency range is usually far greater than that of the maximum frequency deviation.

If a complex waveform is used to frequency modulate a carrier, then sidebands of frequencies are produced as with amplitude modulation but the limit of the upper sideband is not $f_c + f_{max}$ and the limit of the lower sideband is not $f_c - f_{max}$. The upper sidebands exist between f_c and $f_c + f_{max}$, then between $f_c + f_{max}$ and $f_c + 2f_{max}$, then between $f_c + 2f_{max}$ and $f_c + 3f_{max}$ and so on to a maximum determined by the amplitude of the largest component of the modulating signal. A similar situation exists in the lower sidebands (see figure 7.13). As can be seen, a series of sidebands exist, each having an upper or lower limit situated f_{max} hertz from the next. It should be noted that of all sidebands produced, only those closest to the carrier frequency have any significant effect; that is the amplitudes of the side frequencies are considerably reduced as they move further away from the carrier frequency.

In pulse *modulation* (illustrated in figure 7.14) the carrier is transmitted in pulses rather than continuously as in the previous modulation systems considered. These pulses are then modified (modulated) in one of three ways; *pulse amplitude modulation, pulse-width modulation* and *pulse position modulation*. In the first of these the intelligence to be transmitted is used to change the amplitude of the pulses, in the second to change the

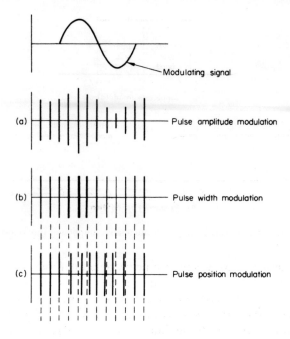

figure 7.14   Pulse modulation

duration of the pulses and in the third to change the position of the pulses relative to their 'normal' unmodulated positions (that is to arrive either before or after the 'normal' times of the unmodulated pulses as the modulating signal goes 'positive' or 'negative'). In all three types the extent of the modulation is governed by the amplitude of the modulating signal.

8 Propagation

Whenever electric current flows in a conductor an electric field and a magnetic field are present in the vicinity of the conductor. If the current is alternating in nature both fields will also alternate, their strengths rising and falling with time. The combination of these fields, which act at right angles to one another (as shown in figure 8.1), produce an electromagnetic disturbance or wave that is capable of energy transference from one point to another. A conductor which is radiating electromagnetic energy in this way is called an *aerial*. The effectiveness of the radiation depends also on the type and length of aerial used; the higher the frequency, the shorter the aerial for best transfer. Mains cables often provide an effective aerial leading usually to undesirable mains radiation. As shorter aerials are obviously more convenient to handle, the higher end of the electromagnetic spectrum is normally used for radio transmission. Other factors taken into consideration in choosing a particular carrier frequency include the method of propagation, which varies with the frequency of the transmitted wave, as described below.

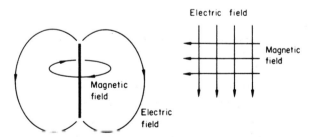

figure 8.1   Electromagnetic field patterns for propagation

Electromagnetic waves are propagated in two ways, one of which can be further divided into two. These are the ground wave and the sky wave; the ground wave consists of two further types, the surface wave and the space wave.

The surface wave travels relatively close to the earth's surface following the curvature of the earth. The space wave follows the direct line between transmitter and receiver so that space wave transmission is referred to as 'line of sight'. If the receiver is below the horizon as far as the transmitter is concerned transmission by space wave alone is not possible. The surface wave component of the ground wave then has more effect.

The sky wave is radiated directly into the upper atmosphere and, depending upon the angle of incidence and the wave frequency, may be reflected back to earth by one or other of the various layers of charged particles that exist above the earth, and are collectively called the ionosphere. The angle of incidence is the angle between the line of propagation of the wave and a line drawn vertically to the surface of the charged layer. The ionosphere itself, which is composed of several layers termed D, E and F, is set up by atmospheric particles that gain or lose charge due to the sun's radiation. The height of the various layers above the earth's surface is determined particularly by the time of day or night and also the time of year, because this in turn is affected by the position of the sun relative to the earth.

Whether or not the sky wave or part of it is reflected is determined by two factors; the angle of incidence (as previously described), and the frequency of the propagated signal. If the angle of incidence is reduced, there comes a point at which the incident rays travel through the ionosphere layer and are lost. This angle of incidence is called the critical angle. The reflective or, to be precise, refractive (bending) properties of the layer (as regards a particular wave) depend to some extent on frequency; and, if the frequency is progressively increased, eventually the incident wave is not reflected but is instead lost in the charged layer. The frequency at which this occurs is called the critical frequency.

At frequencies between 10 kHz and 300 kHz (wavelength 30 km to 1 km), called the long wave band or Band A, transmission is mainly by ground wave, particularly the surface wave.

Attenuation of the surface wave (due to absorbed energy caused by induced surface currents) is least severe over surfaces having large conductivity. Consequently, this band is especially useful for maritime communication. Transmission at these frequencies is, however, subject to interference from atmospheric disturbances and from unsuppressed electrical equipment.

Between 300 kHz and 3 MHz (wavelength 1000 m to 100 m) transmission over short ranges is mainly by ground waves, and over longer ranges (up to 1000 miles) by sky waves. This band is called Band B or the medium wave band.

In Band C, the short wave band, which lies between 3 MHz and 30 MHz, transmission is mainly by sky waves, which are reflected by the ionosphere. The distance covered by the sky wave from transmitter to receiver when the maximum usable frequency is being employed is called the skip distance. Reception is possible a short distance from the transmitter by means of the ground wave, the distance between the final point of the ground wave reception and the point of sky wave reception being called the dead space. See figure 8.2. This band is not as susceptible to interference as bands A and B.

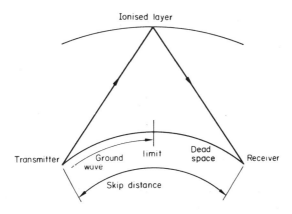

figure 8.2

Band D, 30 MHz to 300 MHz, uses space wave propagation, because the ground wave suffers marked attenuation and the sky wave is not returned in many cases. This band is used for short range television and VHF broadcasting, the receiver and transmitter preferably not being separated by terrain changes (hills etc.) or other obstacles if good transmission is to be obtained.

Band E which lies above 300 MHz is used in radar systems and, as with B and D, line of sight transmission using the space wave is employed.

Aerials

A system to transmit or receive electromagnetically propagated signals is only as efficient as its aerial. Correct matching of transmission lines between aerial and system and the use of high gain amplifiers providing maximum range and quality are no use alone unless the correct aerial is used with the system.

Resonant aerials are so called because they behave in a similar manner to a series resonant or acceptor circuit. Two of the main kinds of resonant aerial are the half-wave dipole and the Marconi quarter-wave aerial illustrated in figure 8.3. The half-wave aerial, which was originally developed by Hertz in 1887, has a length equal to half of the wavelength of the radiated signal frequency, because it is found that with this length maximum resonance effect, and thus maximum effectiveness of radiation is achieved. The quarter-wave aerial, developed from the Hertz design, uses the earth as a conducting plane.

figure 8.3   Marconi quarter-wave aerial

Before considering aerial designs further, it is necessary to define certain terms used to describe their performance. These include *aerial resistance, matching, polarisation* and *polar diagram*.

Aerial resistance is the effective resistance of the aerial used in determining the energy dissipation in the aerial. The aerial resistance is not necessarily the same as the aerial impedance, which includes the inductive and capacitive reactance of the aerial and is determined by the point of connection to the aerial. A centre-connected half-wave dipole, for example, has a resistance and impedance of about 75 Ω; however, if the dipole is end-connected, the resistance remains the same but the impedance may be of the order of 2500 Ω.

It can be shown that for maximum transfer of energy between aerial and line, the line resistance and aerial resistance should be equal. The process of obtaining maximum transfer by arranging this state of affairs is called *matching*.

As also stated, an electromagnetic wave is set up by a magnetic field and an electric field acting at right angles to each other. A horizontally polarised aerial is one that has the associated electric field lying in the horizontal plane; a vertically polarised aerial has the electric field in the vertical plane. Figure 8.1 showed that the electric field plane lies along

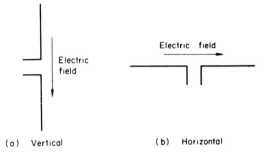

(a) Vertical   (b) Horizontal

figure 8.4   Polarisation

the aerial, so that a vertically mounted aerial is vertically polarised as shown in figure 8.4a. Figure 8.4b shows horizontal polarisation.

If the electric field strength is measured at various points away from the aerial it is found that uniform radiation does not occur in all directions in both the vertical and horizontal planes. If lines are drawn on a diagram joining together all points where the observed field strength is the same, the resultant picture is called a *polar diagram*. A polar diagram for a transmitting aerial shows the direction in which the transmitted energy will be most strongly radiated; for a receiving aerial the polar diagram shows in which direction the aerial should be mounted to obtain best reception. Diagrams for a half-wave dipole are shown in figure 8.5. As can be seen, for a vertically mounted aerial, figure 8.5a, the vertical polar diagram shows zero radiation above and below the aerial and maximum radiation along the centre line of the aerial in the vertical plane. In the horizontal plane the field strength is uniform in all directions for the vertically mounted aerial, as shown by the circular polar diagram.

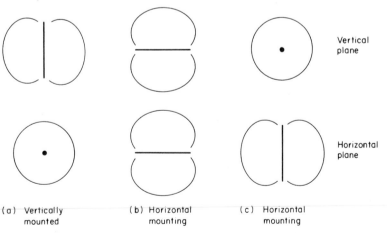

(a) Vertically   (b) Horizontal   (c) Horizontal
      mounted   mounting   mounting

figure 8.5   Polar diagrams

For a horizontally mounted aerial the diagrams are as shown in figure 8.5b and c.

The simple aerials shown are relatively non-directional, for radiation takes place equally in all directions in one plane or the other, depending on the mounting of the aerial. Introduction of additional elements called *parasitic elements* to the simple aerial produce a directional array such as the one shown in figure 8.6. This aerial is a 3 element Yagi aerial with one reflector and one director as shown. The director is in front of the aerial and is shorter than the aerial itself; and the reflector, which is slightly longer, is situated behind the aerial. The resultant polar diagram in the horizontal plane shown in the same figure shows a highly directional aerial as opposed to the simple circle diagram of the aerial alone. Other Yagi arrays are possible with differing numbers of directors etc.

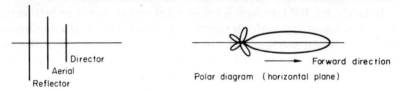

figure 8.6   Three-element Yagi array

For maximum reception of any given signal the factors to be considered are, the frequency (and thus wavelength), the direction of polarisation, and (in certain cases, involving 'line of sight' transmission), the general direction in which the transmitter is situated. The appropriate aerial type, mounting and direction can then be selected.

9 System subunits

Any electronic system may be considered to be made up of a number of subunits, each one containing a combination of active and/or passive components (discussed in chapter 10) designed to process appropriately the electronic signal as it moves through the system. These different processes include generation, amplification, differentiation, integration, modulation, de-modulation and switching. Also all subunits containing active components require electrical energy and this is provided by a further subunit, the power supply. The power supply does not itself generate or process a signal, but without it no generation or processing using active components can take place.

Electronic subunits can be considered initially without detailed knowledge of their circuit arrangements. This chapter is concerned with a general discussion of the overall characteristics of subunits.

Power Supplies

Regardless of whether a.c. or d.c. signals are being transmitted through the system, most circuits containing active components (valves, transistors, diodes etc.) require a power unit that supplies direct current (that is, current flowing in one direction). An obvious method of providing d.c. is by using batteries and a number of systems, especially if they are to be portable (radios, record players etc.) employ them. However, for larger fixed systems requiring rather more current, batteries are not convenient and d.c. power supplies derived from the main or other locally generated electricity supply are used. Mains electricity in most parts of the United Kingdom comes in the form of alternating current and consequently provision must be made within the power supply circuit to change the a.c. to d.c. This process is called *rectification*. There are two kinds of rectification as illustrated in figure 9.1. In one, the sine waveform of the

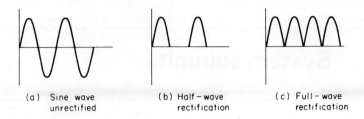

figure 9.1   Sine-wave rectification

supply (figure 9.1a) is modified by removing either the upper or lower half (as shown in figure 9.1b); this process is called *half-wave* rectification and is somewhat wasteful since no use is made of half the supply wave. In the other rectification process, one half of the wave is inverted as shown in figure 9.1c, the resultant current being a series of unidirectional pulses (that is, pulsating d.c.); this process is called *full-wave* rectification.

Both half-wave and full-wave rectification produce a direct voltage which is fluctuating. The waveform of such a voltage can be considered to be made up of a constant direct voltage and an alternating voltage. The direct-voltage component is the one required by the subunits fed by the supply, but the alternating component, or *ripple*, must be removed if the subunits are to operate correctly. This is done by means of a *filter* circuit which allows d.c. to pass virtually unaffected but blocks the flow of a.c. Figure 9.2 shows the waveforms of a poorly filtered output voltage and a well-filtered output voltage. The ideal output voltage would have no alternating component, but it is not possible, however, to completely remove the ripple. It should be noted that the ripple frequency is equal to the input-supply frequency for a half-wave rectifier and equal to twice the input-supply frequency for a full-wave rectifier.

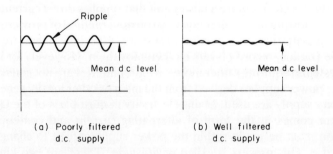

figure 9.2   Effect of filtration

The electrical circuits fed by a power supply are called the *load* on the supply. When a load is connected to a power supply the output voltage changes from the no-load value. An important property of any power supply is its *regulation characteristic* which is a graph showing how the output voltage changes as the load is increased. A poorly regulated supply gives an output voltage that is extremely dependent on load; such supplies are used only when the load is reasonably constant. If the load is likely to vary yet a fairly constant output voltage is required, a regulator circuit or *stabiliser* of some kind must be included between the supply and the load. Voltage regulation curves for a power supply with and without a stabiliser are shown in figure 9.3. Figure 9.4 shows a block diagram of a power supply showing the rectifier, filter and regulator blocks. Most regulators stabilise the supply not only against load variations but also against certain changes in the incoming a.c. supply.

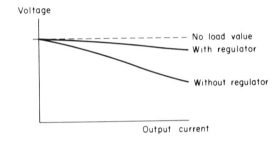

figure 9.3   Power supply regulation curves

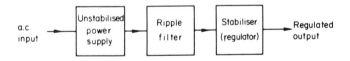

figure 9.4   Stabilised power supply

The system illustrated in figure 9.4 is a simple stabilised power supply. To obtain even better regulation (that is, a more constant output despite varying loads or varying inputs) a more sophisticated system of the kind shown in figure 9.5 may be used. In this type, two d.c. supplies are derived from one a.c. input supply. One d.c. supply is used to provide a constant reference voltage which is independent of load variation and which is proportional to the *desired* output. The other d.c. supply provides the *actual* output. The values of the actual output and the desired output are compared in a subunit called a *comparator*, the output of the comparator being used to control the regulator that determines the *actual* output. Should the actual output change, the difference between

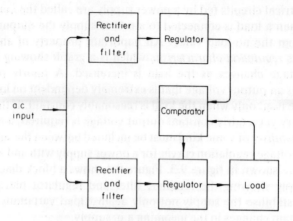

figure 9.5   A more sophisticated stabilised power supply

the new value and the desired value is measured by the comparator which then adjusts the regulator and thereby causes the change in actual output to be reduced to as near zero as possible. The system can be used to provide either a constant voltage or a constant current supply as desired, the reference and output voltages in the latter case being proportional to the desired current and actual current respectively. This process of using the output to control itself is called *feedback*. Feedback of one form or another is used extensively in electronic systems.

The remaining subunit likely to be found in electronic power supplies is the *distributor* which is a circuit so arranged that various levels of voltage can be tapped off from the supply simultaneously. A distributor may be placed before or after the regulator, depending upon which supplies require to be stabilised and which do not.

Amplifiers

One of the most common subunits used for processing electronic signals is the *amplifier*. An amplifier strengthens the signal by increasing the amplitude of the voltage or current or by increasing the power available from the signal. Thus, there are *voltage amplifiers, current amplifiers* and *power amplifiers*. A power amplifier, for example, does not necessarily amplify voltage, but the voltage–current *product* (that is, the power) is greater at the output than at the input of such an amplifier.

As well as being defined in terms of the particular signal property being amplified, amplifiers are also sometimes described by the range of signal frequencies that they are capable of handling. These categories are d.c., a.f., r.f. and wideband. A *d.c. amplifier* handles signals that change at a very slow rate (not more than a few cycles per second); *a.f. amplifiers* handle the audio range of frequencies (up to 20 kHz); and *r.f. amplifiers* handle a narrow band of frequencies within the radio-frequency part of the electromagnetic spectrum. In contrast, a *wideband amplifier* is capable of handling signal frequencies from a few hertz to many millions of hertz; so that this kind of amplifier is used for handling square and other non-sine waves. As indicated in chapter 7, this kind of waveform contains high-frequency harmonics as well as the lower frequency fundamental.

There are seven properties of an amplifier which are of importance and which are used when comparing one circuit with another. These are the input impedance, output impedance, transfer characteristic, gain, frequency response, phase shift and feedback.

The word *impedance* means opposition to the flow of electric current. At zero frequency (that is, direct current) impedance is called *resistance*. The higher the value of impedance or resistance, the greater is the voltage developed across it for a particular current value. Similarly, for a particular value of applied voltage, the higher the value of impedance or resistance, the lower is the value of current that flows.

The *input impedance* of an amplifier or other subunit is the effective impedance between the input terminals as presented to the signal. Effective means that the impedance is not necessarily that of the component(s) seen to be connected across the input, but is a combination of that visible impedance with the impedance of the first active device (transistor, valve etc.) in the circuit. The active device impedance is influenced by a number of factors, including what feedback (if any) is present in the circuit. If internal feedback is present, the load connected to the amplifier output can considerably affect the effective input impedance. The magnitude of the input impedance is an important factor in the *matching* of one subunit to the next. This aspect is now further considered.

The *output impedance* is the *effective* impedance across the output terminals as seen when looking back into the output. As with input impedance, the output impedance is not necessarily that of the component or components connected across the output, but is a combination of that apparent impedance with the impedance of the last active device in the circuit as a whole. With certain active components, notably transistors, the value of the output impedance may be particularly affected by what is connected to the amplifier input.

Matching of an amplifier, or other subunit, to another means arranging the values of the output impedance of the one and the input impedance of the other so that there is as little as possible loss of signal during transfer. Consider figure 9.6 which shows a generator in series with two resistors R_0 and R_i. (A resistor is a passive component having resistance: see chapter 3.) The generator and the resistor R_0 represent the equivalent circuit of an

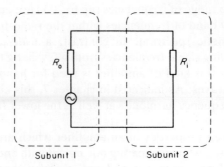

figure 9.6　Input and output resistance

amplifier and its output resistance, or the *effective* circuit as seen by the following subunit; the resistor R_i represents the input resistance of the next subunit. To obtain maximum voltage the input resistance should be high and the output resistance low, because the signal voltage to be transmitted through the system is that across the input resistance. Similarly, if the signal current is to be high the input resistance should be as low as possible; while if signal power is to be high the *maximum power transfer theorem* shows that the input and output resistance should be equal. As can be seen, then, the values of output and input impedances or resistances have a considerable influence on the effectiveness of the matching between one subunit and the next.

The *transfer characteristic* of an amplifier is a graph in which output voltage or current is plotted against input voltage or current (see figure 9.7a). The slope of the transfer characteristic is a measure of how much the amplifier amplifies (that is, of the *gain* of the amplifier): the steeper the slope, the larger is the amplitude of the output for a given input. The gain of an amplifier is determined by the type of active components used and by the pattern in which they are interconnected (this will subsequently be considered in greater detail). When the transfer characteristic is a straight line, the output waveform is a larger version of the input and the input waveshape is faithfully reproduced at the output. A curved transfer characteristic leads to a distorted output (as shown in figure 9.7b). A signal that is distorted when it leaves the amplifier must contain frequencies that the original input did not have, and the effectiveness of the transmission of intelligence is thereby reduced. In practical terms, if the system is a radio receiver, then the output sound will not be a true reproduction of the input, in a television receiver the picture at the screen will not exactly duplicate that seen by the camera.

As already indicated, the *gain* of an amplifier is a measure of its amplification. More precisely, it is the ratio of output signal amplitude to input signal amplitude (voltage or current) or, for a power amplifier, the

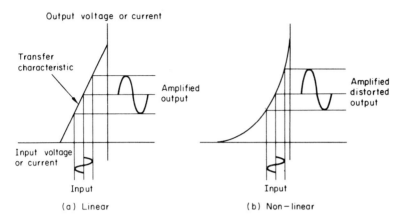

figure 9.7   Transfer characteristics

ratio of output power to input power. Amplifier gain is determined by the individual amplification of the active components used and by the pattern in which they are interconnected, and it varies with frequency. A gain versus frequency graph (known as the *frequency response* curve) may be drawn for any particular amplifier and used to compare its performance with any other amplifier. Typical frequency response curves for audio and radio frequency amplifiers are shown in figure 9.8. The separation between the frequencies at which the gain falls to 0.707 of the maximum for a voltage or current amplifier, or to 0.5 of the maximum for a power amplifier, is called the *bandwidth*. An r.f. amplifier is described as a narrow band amplifier because the bandwidth is small; this is shown in figure 9.8b. Such a small bandwidth represents an amplifier that is very selective and this is precisely what is required in a subunit designed to select and amplify one carrier wave out of many hundreds. The shape of frequency response curves can be considerably changed by the application of *signal feedback* techniques.

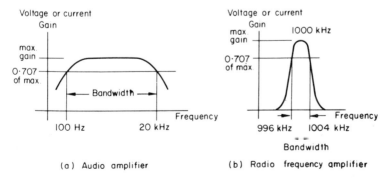

figure 9.8   Amplifier frequency response curves

Phase shift in an amplifier is the amount (if any) by which the output signal is delayed in time with respect to the input signal. Phase shift may be expressed either in terms of time or, more commonly, in 'degrees', where 1 cycle is assumed to occupy 360°. Thus, a phase shift of one quarter of a cycle (that is, where the output maximum occurs one quarter of a cycle after the input maximum) could be expressed as $T/4$ s, where T is the time taken for one cycle (the periodic time), or as 90°.

Phase shift is caused not only by the time taken for the signal to pass through the amplifier circuits but also by certain components in which the impedance is affected by frequency. This is further discussed in chapter 11, which covers both active and passive components. Whether or not the phase shift of an amplifier is important is determined largely by the function of the signal being amplified. In switching systems, for example, in which certain switching actions must take place at a particular time, phase shift may be of the utmost importance and steps may have to be taken to compensate for its effects (see figure 9.9).

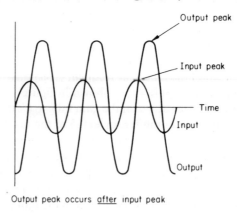

Output peak occurs *after* input peak

figure 9.9   Amplifier phase shift

The feedback in an amplifier is usually expressed in terms of the amount of output signal which is fed back to the input. Often, feedback is deliberately introduced into an amplifier, because it is found that amplifier performance can be improved in this way. In some active devices, particularly transistors, feedback is *inherent* (that is, it occurs *inside* the device) and special precautions must be taken to counteract it in those circumstances where its effects are undesirable. Feedback is an important technique in electronic systems and a more detailed account is given in the next four paragraphs.

Feedback is the process of taking either part or all of an output signal and feeding it back to the input. The feedback signal can be arranged to increase the input signal (the process is then called *positive feedback*) or to

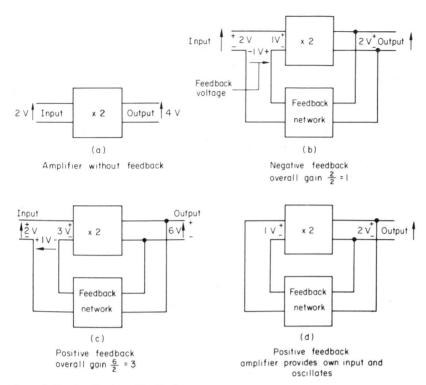

(a)

Amplifier without feedback

(b)

Negative feedback
overall gain $\frac{2}{2} = 1$

(c)

Positive feedback
overall gain $\frac{6}{2} = 3$

(d)

Positive feedback
amplifier provides own input and
oscillates

figure 9.10   Application of feedback

reduce the input signal (the process is then called *negative feedback*). Positive feedback, since it increases the input signal, produces a larger output signal and so the effective gain of the amplifier is increased. Similarly, negative feedback produces a reduced output signal and the effective gain is decreased. This is clarified in figure 9.10, which shows a voltage amplifier with and without feedback. Figure 9.10a shows an amplifier of gain 2 so that an input of 2 V produces an output of 4 V. In figure 9.10b part of the output is taken back to *oppose* part of the input (that is, the feedback signal is in *antiphase* with the input); the initial input of 2 V is now reduced to an effective 1 V giving an output of only 2 V. The effective gain is now 1, as an input of 2 V to the feedback amplifier gives an output of only 2 V. This is *negative* feedback as the effective gain is reduced.

In figure 9.10c part of the output is fed back to assist the input (that is, the feedback signal is *in phase* with the input) so the amplifier receives a signal of 3 V. The output is thus 6 V because the amplifier gain is 2. The overall gain of the feedback amplifier, however, is increased to 3 as an input of 2 V (from the previous subunit) produces an output of 6 V. This feedback is *positive*.

A special case of positive feedback is illustrated in figure 9.10d. Here, part of the output is used to provide *all* the input and the amplifier generates its own signal. The effective gain under these conditions is infinitely high, because an output is generated without an input from a previous subunit. An amplifier connected in this manner is said to *oscillate* and the subunit is called an *oscillator*. Oscillators are widely used in the generation of electronic signals and are considered in the next section. The initial input to set an oscillator in operation is supplied by transient voltage or current surges that arise each time the oscillator is switched on.

The application of positive feedback to an amplifier merely to increase gain is not widely used because of the possibility of the subunit breaking into oscillation. Self-generation of a signal is obviously undesirable in a circuit that is designed to handle a signal already developed elsewhere. Sometimes positive feedback occurs internally within an amplifier and to avoid the possibility of oscillation negative feedback may be deliberately applied. This is not necessarily a bad thing, because although gain is reduced it can remain fairly constant over a much larger frequency range than it is able to do when feedback is not applied. Typical graphs of gain versus frequency with and without feedback are shown in figure 9.11.

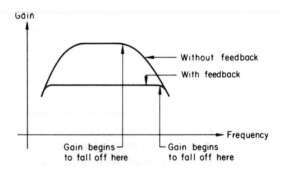

figure 9.11   Effect of negative feedback on frequency response

Oscillators

A subunit that generates a signal is called an oscillator. Oscillators are used to provide a.c. signals from frequencies just above zero to frequencies in the gigahertz region at the top end of the radio-frequency part of the electromagnetic spectrum. They may be described in one of two ways in terms of the type of signal waveform generated: *sinusoidal oscillators* and *relaxation oscillators*. a sinusoidal oscillator generates a signal having a sine waveform; a relaxation oscillator generates a signal that is usually of square waveform.

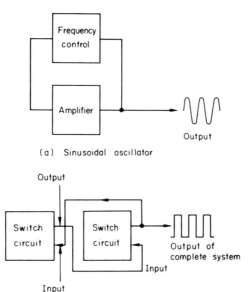

(a) Sinusoidal oscillator

(b) Relaxation oscillator
Output may be taken from either switch circuit

figure 9.12   Oscillators

The sine wave oscillator comprises an amplifier that provides its own input by feedback through some kind of frequency controlling component such as a *tuned circuit* or a *resonant crystal* (as shown in figure 9.12a). These components are considered in more detail in subsequent chapters. A relaxation oscillator comprises two switching circuits so interconnected that the output of each one in turn switches the other. The switching process is continuous so that the output waveform is that of a d.c. level being continually interrupted as shown in figure 9.12b.

Sinusoidal oscillators are widely used in radio communication for carrier generation and in many of the test instruments used with such systems. Relaxation oscillators are used as pulse generators in television and radar systems, in digital systems (computers and logic circuits) and in test instruments. Detailed circuits of both types of oscillator are presented in *Electronics Servicing Part II: Core Studies.*

Modulators, Demodulators and Detectors

Modulation was discussed in chapter 7. As explained there, it is a process by which intelligence to be conveyed is impressed upon a higher frequency carrier signal, the reason for the use of the process being that it is easier to transmit electromagnetic waves at frequencies higher than those of the intelligence. A subunit that carries out the modulating process is called a *modulator* and such subunits are found at the transmitting end of the system. A modulator has two inputs and one output: the inputs are the signal that is to be modulated and the signal that is to do the modulating; the output is the resultant modulated carrier.

Once received, the intelligence signal must be extracted from the carrier. A subunit to carry out this process is called a *demodulator* or *detector* and it has one input and one output, and the output carries the intelligence. The carrier wave is usually discarded by suitable filters, because once the signal is received the carrier has completed its task. A demodulator used to extract the intelligence from a frequency-modulated carrier is often given the name *discriminator*.

As was stated earlier, a radio frequency amplifier is designed to amplify a narrow band of frequencies centred about the carrier frequency that it is desired to receive. In order to change the required carrier (that is, to select another radio station) it must be possible to change the centre frequency to that of the new carrier. This process of moving the frequency response curve is called *tuning*. For weak incoming signals more than one r.f. amplifier may be necessary to strengthen the signal sufficiently before detection. As will subsequently be explained in more detail, it is not convenient to have a series of tunable amplifiers in one receiver. Consequently a process called *frequency conversion* is used, in which the carrier signal frequency (or carrier signal centre frequency for frequency modulation) is converted to a constant radio frequency regardless of the value of the incoming signal frequency. This then means that subsequent r.f. amplifiers need not be tunable but can be designed to have a bandwidth centred about a single constant frequency. This constant frequency is called the *intermediate frequency* (i.f.). The r.f. amplifiers included to handle the i.f. are called i.f. amplifiers to distinguish them from the tunable r.f. types.

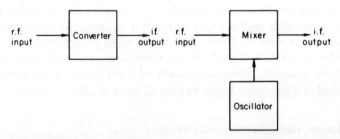

figure 9.13   Mixing and frequency conversion

Frequency conversion, or *mixing*, uses a similar process to modulation, in that the incoming modulated r.f. is mixed with a second signal derived from an oscillator, to give an output waveform made up of a number of component frequencies, from which frequency-selective circuits are then used to extract the i.f. The i.f. signal is itself modulated in exactly the same way as the original r.f. but the carrier frequency is now different.

The name *converter* is usually given to a subunit that converts the r.f. to i.f. and contains its own oscillator. The name *mixer* is used for a subunit that requires a separate oscillator (see figure 9.13).

Differentiators and Integrators

Differentiation is the name given to the mathematical process of determining the *rate of change* of a varying quantity. If a graph of the varying quantity is drawn the rate of change is given by the *slope* of the graph. For example, consider a voltage increasing with time at a constant rate as shown in figure 9.14a. This waveform is called a *ramp function* because of its shape. As the rate of change of a ramp function is constant, a graph of the rate of change looks like the one shown in figure 9.14b; that is, a straight line at a constant height above the time axis. If this straight line represented a voltage we would say that its waveform was that of a differentiated ramp function.

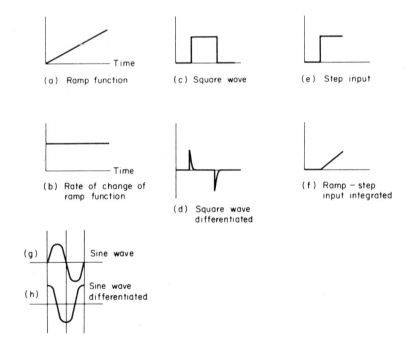

(a) Ramp function

(c) Square wave

(e) Step input

(b) Rate of change of ramp function

(d) Square wave differentiated

(f) Ramp – step input integrated

(g) Sine wave

(h) Sine wave differentiated

figure 9.14   Integration and differentiation

The opposite process to that of differentiation is called *integration*. Thus if the straight line waveform of figure 9.14b is integrated it gives the ramp function of figure 9.14a. With this process we are finding the waveform whose rate of change graph is the same shape as the waveform to be integrated. To further explain these two processes consider some typical signal waveforms as shown in the remainder of figure 9.14. The waveform in figure 9.14c is a square wave. The rate of change of such a wave is rapid and positive when the square wave rises, zero over the portion that the wave is constant, and rapid and negative when the square wave falls. If a square wave is differentiated the resultant waveform would then be that shown in figure 9.14d; that is, a series of alternate positive and negative pulses. Figure 9.14e shows the first part of a square wave: the sudden rise to a positive constant level. Such a waveshape is called a *step-function*, from its appearance. The rate of change of a ramp function is constant so that if a step-function is integrated the resultant wave is a ramp function as shown in figure 9.14f. Finally, a sine wave is shown in figure 9.14g. The rate of change of a sine wave is a positive maximum as it goes from negative to positive through zero, is zero when the wave reaches a maximum and is a negative maximum as the sine wave goes from positive to negative through zero. The graph of the rate of change is, in fact, itself a sine wave but displaced a quarter of a cycle to the left (that is, a sine wave shifted *in phase* by the time taken for a quarter of a cycle). If a sine wave is differentiated the resultant waveshape is thus another sine wave but displaced in phase to the left. Such a phase displacement is called a *lead*; that is, the differentiated wave *leads* the original wave by a quarter-cycle (see figure 9.14h). Conversely, an integrated sine wave will be phaseshifted a quarter-cycle to the *right*; that is, will *lag* the original wave.

Integrating and differentiating circuits, called integrators and differentiators respectively, may contain all passive components or a combination of active and passive components. Under certain conditions certain kinds of amplifier may be used to carry out these processes.

Limiters, Clippers and Clamps

Sometimes in certain electronic systems it is necessary to limit the signal amplitude to a particular level. A subunit that does this is called a *limiter*. One example of the use of a limiter is in an *FM* receiving system. As was described in chapter 7, frequency modulation of a signal means using the intelligence to be transmitted to modify the frequency of the carrier signal. The detector circuit (discriminator) used to extract the intelligence from the carrier as it proceeds through the receiver may well be sensitive to amplitude changes as well as frequency changes, in which case any amplitude change might be taken as part of the intelligence signal.

This would produce distortion of the intelligence signal. An amplitude limiter inserted before the discriminator removes the problem.

Clipping is a similar process to limiting and often the two words are used to mean the same thing. Clipping consists of removing a section or 'slice' of a waveform as shown in figure 9.15. One use of the process is to obtain an approximate square wave from an original sine wave by taking the 'slice' shown in the figure and straightening the rise and fall parts of the signal. If there is a difference in the two processes, clipping and limiting, it is probably that a limiter retains the original waveform whilst controlling its amplitude; whereas a clipper, which also controls amplitude, does not necessarily retain the original waveform.

Clamping a signal either sets the d.c. level about which the signal waveform alternates, or prevents a signal from rising above a particular level. In the latter case clamping is similar to limiting or clipping, but only on one half of the signal waveform.

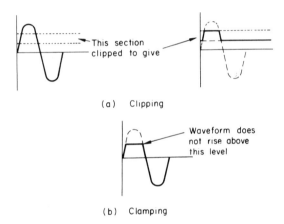

(a)   Clipping

(b)   Clamping

figure 9.15   Limiters, clippers and clamps

Logic Gates

An electronic *gate* is a subunit that allows a signal to pass under certain conditions. If these conditions are not satisfied the gate remains closed and the signal cannot proceed. The dictionary defines the word 'logic' as a 'connection or outcome of events' and it is in this sense that the word is applied to electronic gates; the connection of events being that if certain conditions are satisfied (that is, if certain events take place) then a further event (that of the signal proceeding) follows logically. The field of logic in electronic systems is very broad and logic theory is being applied more and more to all kinds of control, measurement and calculating systems. Logic theory is, of course, the basis of what is probably the best known electronic system, the *digital computer*.

Logic circuitry in general comes under the branch of electronics known as *digital applications* as opposed to *linear* or *analogue applications*. Digital systems use signals which are of pulse form, that is, a series of short-duration square waves; analogue systems use signals of, or derived from, the familiar sine waveform. Amplifiers used in digital systems must be *wideband* amplifiers because of the many harmonics that make up a square wave.

Logic signals have a unique characteristic in that the signal is a d.c. signal that is always at one of only two levels. These levels are described as 'high' and 'low', or referred to numerically as '1' and '0'. If the low level (that is, the less positive level) is denoted by 0 and the high level by 1, *positive logic* is said to be in use. If 1 and 0 denote the low and high levels respectively, *negative logic* is said to be in use. The main advantage of using only two levels is that they are easily distinguished from one another (particularly if, as is usual, they are generated by switching a circuit either 'on' or 'off'). There can be little confusion in deciding whether or not a circuit is conducting. If several levels are used, however, confusion can arise due to the *drift* of a d.c. signal when passing through a system. Severe drift could cause a signal to change from one level to the next if several levels close to one another were used. It is, however, unlikely that drift would change a level corresponding to the 'on' condition to that corresponding to the 'off' condition.

There are *three* types of logical function performed by electronic gates: these are known as the AND, OR and NOT functions. Gates to carry out these functions are called AND gates, OR gates and inverters (or negaters) respectively. In addition, gates carrying out the NOT-AND and NOT-OR functions are called NAND gates and NOR gates respectively. All gates have one output and one or more inputs. The symbols for the different gates are shown in figure 9.16. The functions are defined as follows:

The AND function is performed when a gate prevents the output signal attaining logic level 1 until *all* the inputs are at logic level 1. The OR function is performed when a gate allows the output to attain logic level 1 if at least *one* input is at logic level 1. The NOT function is performed if the gate output is at the opposite level compared to the input (that is, if the output is 1 when the input is 0, or vice versa).

To clarify these definitions consider a two input AND gate. The possible combinations of inputs are as follows:

Input 1	Input 2	Output
0	0	0
0	1	0
1	0	0
1	1	1

This type of table showing input and output levels is called a *truth table*. The truth table shows that the output is 1 only when *both* inputs are 1 (see figure 9.16). The truth table for a two input OR gate is as follows:

Input 1	Input 2	Output
0	0	0
0	1	1
1	0	1
1	1	1

Here the output is 1 when one (or more) input is 1 (see figure 9.16). The last line is the same as that of the two-input AND gate truth table above: the OR function includes the AND function. An OR gate that does *not* include the AND functions is called an *exclusive* OR gate. The truth table for such a gate having two inputs is as follows:

Input 1	Input 2	Output
0	0	0
0	1	1
1	0	1
1	1	0

The symbol for an exclusive OR gate is shown in figure 9.16. The truth table for an inverter is as follows:

Input	Output
0	1
1	0

A gate which carries out the AND function and then the NOT function is called a NAND gate. The truth table for two inputs is as follows:

Input 1	Input 2	AND	NAND Output
0	0	0	1
0	1	0	1
1	0	0	1
1	1	1	0

The output is the AND function negated (or *inverted*). Another name for the inversion process is *complementation*, so we can also say that the output is the *complemented* AND function.

A gate that performs the OR function and complements the result is called a NOR gate. The truth table for a two-input NOR gate is as follows:

Input 1	Input 2	OR	NOR Output
0	0	0	1
0	1	1	0
1	0	1	0
1	1	1	0

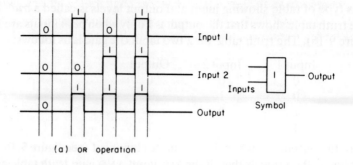

(a) OR operation

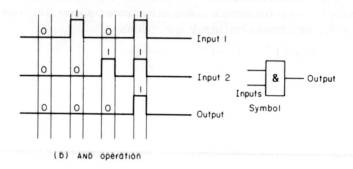

(b) AND operation

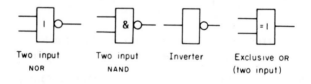

| Two input NOR | Two input NAND | Inverter | Exclusive OR (two input) |

(c) Other symbols

figure 9.16 Logic gate operation and symbols. Note that in parts (a) and (b) the most positive level is taken as logic level 1 (that is positive logic is used)

Changing the type of logic (that is, positive to negative, or vice versa) for a particular gate changes the function of that gate. Consider a gate that has the following truth table:

Input 1	Input 2	Output
Low	Low	High
Low	High	High
High	Low	High
High	High	Low

This means that if the signal level at inputs 1 and 2 is low, the output signal level is high and remains high unless *both* input signal levels are high. Positive logic denotes high by 1 and low by 0, so the truth table becomes as follows

Input 1	Input 2	Output
0	0	1
0	1	1
1	0	1
1	1	0

which is the truth table of a two-input NAND gate. Negative logic denotes high by 0 and low by 1, so the truth table becomes as follows

Input 1	Input 2	Output
1	1	0
1	0	0
0	1	0
0	0	1

which is the truth table of a two-input NOR gate. It can be seen therefore that the same gate is capable of a dual function, and this is the reason why most manufacturers describe their gates as NAND/NOR.

NAND/NOR gates may be used in numerous ways to make up complete systems capable of counting, measuring and computing. When setting up a complete system the truth table for the overall system is drawn up; this then gives the required output signal or signals for the various combinations of inputs. From the truth table equations are obtained using a form of mathematics known as *Boolean Algebra*. These equations are then simplified and the result gives the numbers and types of gate required and indicates how they must be interconnected to perform the desired overall function. Other subunits used in conjunction with the basic gates include flip-flops and multivibrators.

Flip-flops, Multivibrators and Memories

A *flip-flop*, or *bistable*, is a subunit having one (or more) outputs capable of holding a signal in either one of two possible states: the output can be at level 1 or 0 and hold this level indefinitely. On the receipt of an input trigger pulse, the output changes state to the complement of the previous state (that is, from 1 to 0, or vice versa) as in figure 9.17a. A flip-flop is thus capable of *storing* a signal level and can thereby act as a *memory*.

A *monostable* subunit has only one stable state. An input pulse will change the output to the complement but it will not remain in that state; after a certain time the output returns to the state that existed before receipt of the input pulse. The output waveform is shown in figure 9.17b.

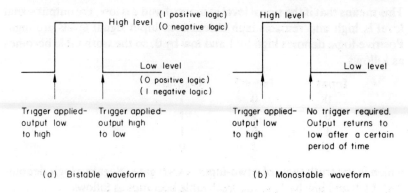

(a) Bistable waveform   (b) Monostable waveform

figure 9.17   Switching circuit waveforms

An *astable multivibrator* or *pulse generator* is a form of *relaxation oscillator* (considered earlier). This kind of subunit is used to generate the input signals to close and open the various gates at the required time and in the required order, and it produces a pulse waveform of the type shown in figure 9.12b.

10   System block diagrams

Chapter 7 described the various kinds of electronic signal and in chapter 9 a number of the more commonly used subunits for signal processing were considered. This chapter is concerned with joining together the various subunits to form complete systems.

Radio Systems

Figure 10.1 shows a block diagram of a simple transmitting system for the electromagnetic propagation of audio signals. It consists of four subunits and does not include the power supply which provides direct current to all subunits. The master oscillator generates the signal carrier frequency which is then transferred to the modulator. The audio frequency intelligence signal is derived from a transducer (a microphone, record player pick-up etc.), amplified in an audio frequency amplifier and then transferred to the modulator. The a.f. signal is impressed upon the r.f. carrier using one of the available forms of modulation (AM, FM etc.) and the modulated carrier is then again amplified in the r.f. power and amplifier to provide a high output to the aerial system.

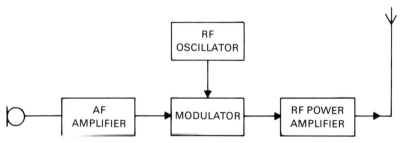

figure 10.1   Transmitter

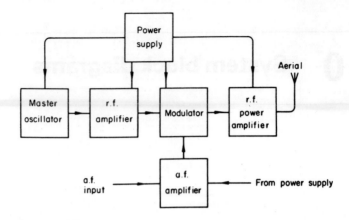

figure 10.2   Simple transmitter system

Figure 10.2 shows a similar system with an additional two subunits, the power supply which is connected to the remaining subunits and an additional radio frequency amplifier between the master oscillator and the modulator. This latter subunit is used to amplify the oscillator signal before it is modulated by the audio frequency signal.

Figure 10.3 again shows a similar system for AM transmission and this will be seen to include a buffer amplifier between oscillator and r.f. amplifier. This subunit helps to reduce frequency drift of the carrier, an essential requirement in an AM system. The r.f. amplifier shown here is indicated as a frequency doubler. This subunit gives an output frequency equal to twice the input frequency. The use of such a subunit allows the use of frequency stability control devices such as piezoelectric crystals (discussed later) at a much lower frequency than that used for propagation.

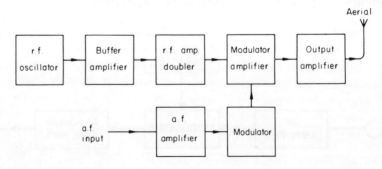

figure 10.3   AM transmitter

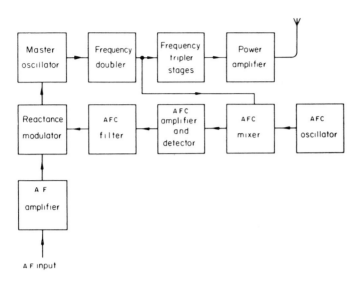

figure 10.4   FM transmitter with AFC

The block diagram of an FM transmitter using a reactance modulator is shown in figure 10.4. This system also employs automatic frequency control (AFC). The master oscillator produces the carrier frequency as before. This output is then progressively doubled and tripled before being fed to the aerial system via a power amplifier. The master oscillator frequency is controlled by a reactance modulator, which puts a variable capacitance across the oscillator.

The value of this capacitance is in turn controlled by the audio amplifier that provides the audio intelligence. As stated in chapter 9, the carrier frequency then changes in accordance with the variation in amplitude of the audio signal. In the absence of a signal the carrier should return to the centre frequency. Sometimes this centre frequency drifts and this in turn moves the entire variation (that is, both sidebands) further along the frequency spectrum. This would, of course, create a problem in reception by an FM receiver tuned to receive an FM signal swinging about the given centre frequency. To prevent drift in centre frequency an AFC subsystem consisting of a second oscillator, mixer, amplifier, detector (discriminator) and filter is employed. A signal for this subsystem is derived from an appropriate point in the main system (the output of the doubler, as shown in figure 10.4) and mixed with the fixed output of the AFC oscillator. (This oscillator is usually crystal controlled.) The mixer output is then amplified and fed to a detector that is tuned to the frequency which should be present when the correct carrier centre frequency is mixed with the AFC oscillator output. The mixer output is then fed via a filter to the modulator. If the carrier centre frequency changes, the mixer output changes and the detector produces an output voltage that controls the modulator, thereby returning the master oscillator to the correct frequency. When the carrier centre frequency is correct the discriminator produces zero output and no control voltage is applied to the modulator. It may not be immediately apparent how this system is able to differentiate between normal change in carrier frequency due to the audio signal and drift in carrier frequency due to the centre frequency shifting. Fortunately, these can be distinguished, since frequency drift is a slower process than normal modulation change. This is the purpose of the filter between discriminator and modulator, for it allows slow changes to take effect but does not respond to fast changes. Thus, the AFC system does not cancel the normal modulation but controls the master oscillator only when the carrier frequency drifts. A similar system may be employed in an FM receiver to ensure it stays tuned to the correct centre frequency.

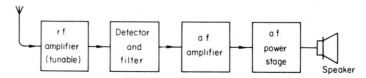

figure 10.5   TRF receiver

A simple tuned radio frequency (TRF) audio frequency receiver is shown in figure 10.5. The signal is selected by tuning the radio frequency amplifier to give maximum gain at the desired frequency. The amplified carrier is then passed to a detector/filter circuit which extracts the audio signal and removes the r.f. component of the carrier. The audio signal is then further amplified to a level sufficient to drive a loudspeaker or other output transducer. This simple receiver suffers from poor selectivity (that is, it is difficult to select a single carrier frequency free from interference by others) and sensitivity (that is, weak signals may be insufficiently amplified). The reason for this lies mainly in the fact that in a TRF system all r.f. stages must be tunable. In a supersonic heterodyne receiver (or superhet for short) only the input r.f. stage, local oscillator and mixer need be continuously tunable. Such a system for AM reception is shown in figure 10.6.

In the superhet system, the first r.f. stage is tuned to the desired carrier frequency, and a separate local oscillator stage is adjusted simultaneously to provide an output at a frequency that exceeds the frequency of the desired signal by a fixed amount called the intermediate frequency. Both the input r.f. and the local oscillator output are fed to a mixer stage which produces an output at a frequency that is equal to the difference between them; which is, of course, the intermediate frequency (i.f.). The i.f. modulation is exactly the same as that of the original r.f. input. The i.f.

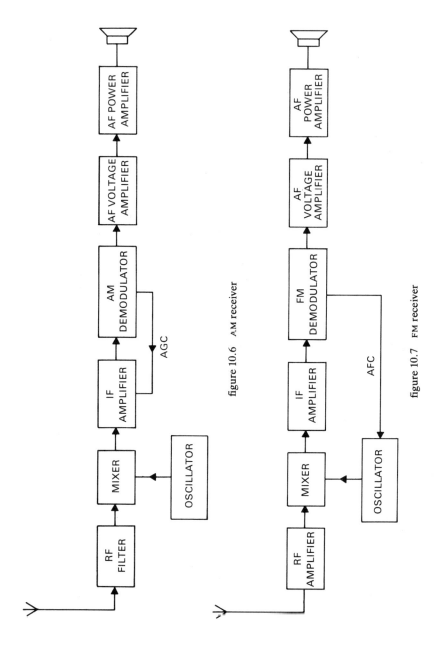

figure 10.6   AM receiver

figure 10.7   FM receiver

signal is now passed through a series of amplifiers set to give maximum gain at the intermediate frequency. These amplifiers are pretuned to the i.f. frequency, as whatever the input frequency from the aerial, the local oscillator is simultaneously adjusted to give an i.f. that is always the same. Cascading the i.f. amplifiers in the manner shown gives a very selective high gain system. From the i.f. amplifiers the signal is detected and filtered, and the a.f. is amplified in the same way as in the TRF system.

The line between the demodulator and i.f. amplifier marked AGC shows the signal path of a feedback signal taken from the demodulator and used to control the gain (amplification) of the i.f. amplifier. Thus when smaller input signals are received the gain is arranged to increase and when larger input signals are received the gain is reduced. In this way a relatively constant output at the speaker is obtained despite fluctuating input signals. The method is called *automatic gain control*, hence the letters AGC.

The use of the superhet principle is not confined to AM reception. A block diagram for a superhet FM receiver is shown in figure 10.7. The arrangement is similar to figure 10.6. If automatic frequency control (AFC) is incorporated in the system as shown the discriminator output is used to provide additional control over the local oscillator. Thus, if the local oscillator drifts, the (relatively slow) change in the intermediate centre frequency is detected by the discriminator, and the local oscillator is returned to the correct setting for the desired input (that is to a frequency that exceeds the input centre frequency by the value of the i.f.).

Television Systems

Television system signals are of necessity more complex than radio system signals. In a monochrome (black and white) television receiver, the electron beam in the picture tube—a form of cathode ray tube (see chapter 13)—moves rapidly from left to right across the screen, returns to the left-hand side at a point slightly lower down, then repeats the movement across to the right-hand side. In this way the beam creates a series of lines on the screen (405 or 625 in the UK). As the beam 'scans' the screen in this way the illuminating effect of the beam is increased or decreased as required (by the picture signal received from the television camera) in accordance with the light distribution on the scene that the camera is simultaneously viewing. The spot of light on the screen thus appears brighter or darker as the beam moves spot by spot over the full area of the screen, and the picture that is built up therefore duplicates that seen by the camera. The camera is basically similar, in that a battery of light sensitive 'cells' pointing at the scene to be viewed is scanned by a beam of electrons and the electric charge distribution on these 'cells', which is determined by the light distribution on the scene, is monitored by the beam to provide the picture signal. It is essential that the camera beam scans in exactly the same way and is in the same relative position at the same time, otherwise the top left-hand corner of the view seen by the camera, for example, might appear at the bottom right-hand corner—or indeed anywhere—on the receiver screen.

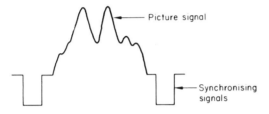

(a) Monochrome TV signal showing line sync pulses

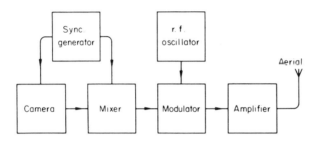

(b) Simplified monochrome TV transmitter

figure 10.8   Monochrome TV transmission

To achieve this, synchronising (sync.) signals are sent from the transmitter to the receiver so that the varying voltage used to drive the spot horizontally (the line timebase voltage) and the varying voltage used to drive the spot vertically (the field timebase voltage) are at the same relative values at the same time. (Note: the term 'field' has now displaced the older term 'frame'.) The composite video signal is thus composed of field and line synchronising signals and the picture signal (see figure 10.8a). The sound associated with the picture is transmitted separately by a conventional FM system at a different carrier frequency to that of the video signal.

The simplified diagrams of a television transmitter and receiver for monochrome signals are shown in figures 10.8b and 10.9, respectively. In the transmitter the synchronising signals are fed from an appropriate generator to the camera and to a mixer unit. In this unit the picture signal determined by the light distribution seen by the camera is mixed with the synchronising signals to give the composite video signal. This signal then modulates the r.f. carrier and is transmitted in the same way as a radio signal. The receiver selects the required signal, detects (demodulates) it to produce the video signal, and this is then split up into the three parts, picture signal, field synchronising signal and line synchronising signal. The picture signal is fed to the electron gun of the picture tube (chapter 13) while the field and line sync. signals are fed to the tube vertical and horizontal deflection systems respectively.

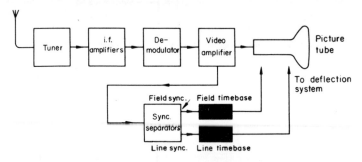

figure 10.9   Monochrome TV receiver simplified block diagram

A colour television system uses a similar principle to the monochrome, in that the picture and sync. components make up the composite signal; but, as might be expected, the system is more complex, for three colour signals are involved instead of a single monochrome signal. The colours are red, blue and green, and from these a picture of all colours normally seen can be constructed. The picture tube commonly used has a complex screen composed of millions of dots of suitable material which glow red, blue or green when struck by an electron beam. The electron gun in these tubes produces three beams, one per colour. By using a device called a 'shadowmask', it is arranged that each beam strikes only its own series of dots (that is, the beam controlled by the 'red signal' strikes only the dots which glow red, and so on).

The composite signal in a colour TV system can be considered to be made up of four parts: the *luminance component*; the *colour component*; the *picture synchronising component* (made up in turn of line and field signals); and the *colour synchronising signal component*.

A simplified block diagram of a colour receiver is shown in figure 10.10. The luminance signal controls the brightness of each colour dis-

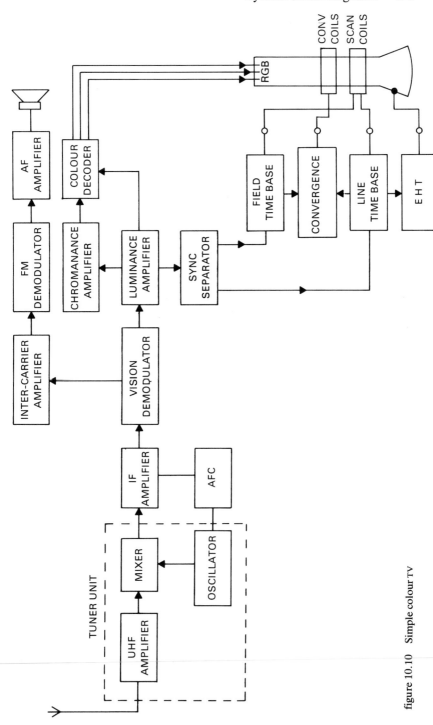

figure 10.10   Simple colour TV

played on the screen and after amplification (controlled by 'brightness' and 'contrast' controls at the receiver) is passed to the picture tube cathodes. The colour or *chrominance* signal is amplified (controlled by the receiver 'saturation' control) and then passed first to a colour demodulator and subsequently to a colour decoder which separates the signal into the separate colour signals for red, blue and green. These are then applied to the appropriate grids in the picture tube. The colour sync. signal acts as an advance warning that a colour signal is coming through, and is contained in a stream of colour 'bursts'. This signal is amplified and then fed to a local oscillator which together with the colour demodulator helps to separate the colour signal into its three component parts. The frequency of the colour local oscillator is synchronised with the colour sub-carrier frequency of the original transmitted signal by the colour discriminator situated between the colour burst amplifier and the colour local oscillator. The picture synchronising signal contains line and field sync. signals as in the monochrome receiver. In addition to being fed to the appropriate parts of the tube deflection systems, they are also fed to a convergence circuit which ensures that the three beams are maintained at the correct spacing relative to each other as the beams move over the screen.

The Cathode Ray Oscilloscope

The cathode ray oscilloscope or c.r.o. is an electronic instrument using a cathode ray tube (described in chapter 13) to display voltage waveforms on a screen. An electron beam from the cathode of the tube is shaped and focused on to the screen, the exact position of striking the screen being determined by the beam deflection system. Television receivers use electromagnetic deflection methods (coils on the neck of the tube) but in the c.r.o. deflection is achieved electrostatically by passing the beam through a set of horizontally-deflecting or X-plates and through a set of vertically-deflecting or Y-plates. When a voltage is applied across these plates the beam moves away from the negative plate and towards the positive plate. It is not allowed to touch the plates since if it does so nothing will be seen on the screen. By positioning the plates correctly the direction of deflection (horizontal or vertical) may be determined accordingly.

A voltage waveform is a graph plotting voltage against time, voltage usually being plotted vertically (in the Y-direction) and time being plotted horizontally (in the X-direction).

When a c.r.o. is used to display the waveform of a signal the signal is applied to the Y-plates and a second time-varying voltage is applied to the X-plates. This voltage is called the *timebase voltage*.

The timebase voltage is used to move the beam from left to right across the screen repeatedly, its own waveform normally being that shown in figure 10.11. This waveform is called a ramp function, from its appearance, and causes the right-hand X-plate to become progressively more positive with respect to the left-hand X-plate, thus drawing the beam from left to right across the screen. Before the beam actually touches the right-hand plate its potential is reduced rapidly to zero and the beam 'flies back' to the left. The right-hand plate is now made increasingly positive

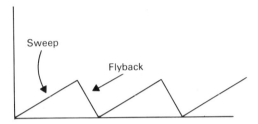

figure 10.11   Time base waveform

again and the process is repeated. If no signal is applied to the Y-plates, the resultant trace on the screen is a horizontal line corresponding to the time-axis of a graph. If a signal is applied to the Y-plates, when the timebase voltage is not applied to the X-plates, the beam is deflected up and down between the Y-plates, the deflection in one direction being determined by the size of the positive-going peaks of the waveform and in the other direction by the size of the negative-going peaks, and the result is a vertical straight line.

If both signal and timebase voltages are applied to the Y-plates and X-plates respectively the beam is deflected vertically by the signal and horizontally by the time base and the spot at the screen traces out a graph of signal voltage (vertically) plotted against time (horizontally). The result is the signal voltage waveform.

The speed at which the beam is drawn horizontally across the screen is important in determining what appears on the screen. If the beam moves at a speed such that the time taken to cross the screen is greater than the periodic time of the signal (the time per cycle) more than one cycle of the signal waveform will be shown, if the speed is such that the time taken is less than the periodic time, less than one cycle appears.

If the time taken is equal to the periodic time then exactly one cycle of signal waveform will be shown. Normally in a c.r.o. it is possible to change the speed of horizontal travel by altering the frequency of the timebase voltage.

If the point on the signal waveform at which the beam begins its movement from left to right (that is, the furthest point to the left at which the spot appears) is not the same each time the beam commences its horizontal sweep across the screen, each horizontal movement will produce a different picture and the waveform will appear to move horizontally. If the starting point is further along to the right each time, the waveform appears to move backwards and, if the starting point is further along to the left, the waveform appears to move forwards. This is obviously confusing if the waveform is to be studied so an additional synchronisation circuit is included to ensure that at the time the spot is just about to commence moving to the right, the signal voltage has reached exactly the same spot on the signal waveform. The screen picture is then stationary. The c.r.o. normally has a 'sync. adjust' control to obtain a stationary trace.

The block diagram of a basic c.r.o. is shown in figure 10.12 omitting the power supply. The input signal is applied via an attenuator (to reduce the signal if it is too large) and an amplifier (to increase the signal if it is too small) to the Y-plates. The signal is also used to trigger the synchronising circuit which in turn controls the timebase unit for the X-plates. The timebase unit is then connected via the X input amplifier to the X-plates. Sometimes it is desired not to display graph-type waveforms and the X-plates may be used as an input for a signal rather than as an input for a timebase voltage. (One example is the comparison of signals, one at the Y-plates, one at the X-plates. The resulting traces, called Lissajou's figures, are used in frequency measurement among other things.) In this case the X-amplifier is disconnected from the timebase unit and connected to an external X input terminal via the switch shown.

Speed Control System

A typical form of electronic speed control system is shown in figure 10.13. Many industrial applications of d.c. motors require the speed to be finely controlled and held within certain specified limits. The use of feedback as shown in the figure achieves this fine control.

The system contains five units: a set-speed control unit, a differential amplifier, d.c. power amplifier, d.c. generator and the d.c. motor which is to be controlled. The d.c. motor speed is determined by the voltage applied to it and this voltage is derived from the d.c. power amplifier, which amplifies a small input voltage to the level required to change the driving voltage of the controlled machine. The input to the d.c. power amplifier, obtained from the differential amplifier, is the difference between two voltages, one being the 'set speed', voltage adjusted by the operator to correspond to the required speed, the other being the 'actual speed', voltage derived from a generator mechanically coupled to the controlled motor and therefore proportional to the actual speed at which the motor-generator set is running.

The method of control is as follows. The set-speed control is adjusted

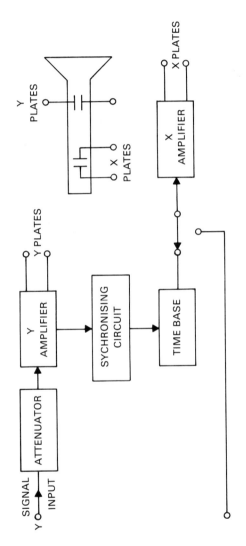

figure 10.12   Cathode ray oscilloscope

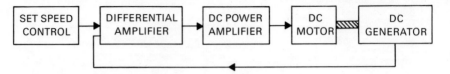

figure 10.13   Speed control system

to give an output voltage equal to the output voltage of the generator when the motor-generator set is running at the desired speed. The motor is started and runs up to speed, the generator output voltage increasing proportionally as the speed increases. The differential amplifier compares the set-speed voltage input with the actual-speed voltage input, derived from the generator, and the difference voltage is amplified in the d.c. power amplifier and applied to the motor. As long as the set-speed signal exceeds the actual-speed signal, the difference voltage applied to the motor causes the speed to carry on increasing. If the actual-speed signal exceeds the set-speed signal, because the motor has over run and is running faster than required, the difference voltage, which is now of opposite polarity causes the motor to slow down. Steady state is achieved when the set-speed and actual-speed signals are equal and there is zero signal to the d.c. power amplifier. If the motor speed changes in either direction, up or down, the system provides a controlling 'adjust-speed' signal at the input to the d.c. power amplifier.

Magnetic Tape Recorder

In the magnetic tape recorder, which is used extensively for recording both audio and video signals, a plastic tape coated with powdered red iron oxide or other magnetic material is subjected to a magnetic field which fluctuates according to the strength and frequency of the signal being recorded. The magnetic coating becomes magnetised in a pattern which is unique to the signal and retains this pattern once the signal is removed as a permanent record of the signal. When playback of the recording is required the magnetised tape is used to set up a fluctuating magnetic field, identical to that which caused the magnetisation, which is reconverted to an electrical signal suitable for application to a loudspeaker or headphone.

The essential units in a magnetic tape recorder system used for audio signals are shown in the block diagram of figure 10.14. These are a record amplifier, record/playback head, erase head, playback amplifier, a.f. amplifier and loudspeaker. The record head, normally also used for playback, consists of a coil wound around a magnetic core which has a gap where the tape moves across its surface. The tape is transferred from one reel to another using a system of drive motors which maintain a constant

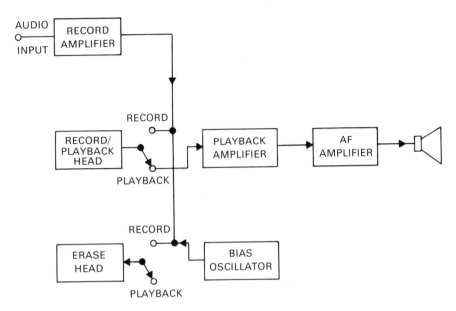

figure 10.14   Magnetic tape recorder

tape speed, playback when required being at the same speed as that used during recording. For high fidelity and best high frequency response (that is, for the playback to be as accurate as possible a reproduction of the recorded signal) a tape speed of 38.1 cm/s (15 inches/second) is found to be ideal but sub-multiples of this are used for lower degrees of fidelity and a frequency response which is not as good but is acceptable for certain purposes (e.g. speech only). These speeds are 19.05 cm/s, 9.525 cm/s and 4.76 cm/s (7.5 inches/second, 3.75 inches/second and 1.875 inches/second respectively). The erase head is of similar construction to the record/ playback head but carries an alternating current at a frequency above sound, that is in excess of 20 kHz, which produced a strong rapidly changing magnetic field that erases the pattern of magnetisation on the tape as it passes the head. On 'record' the tape passes the erase head just before the record head so that the erasure takes place prior to the new recording.

The amplifier units carry out the function their name implies, the record amplifier enlarging the input signal to a level required to produce the correct magnitude of recording magnetic field and the playback and a.f. amplifiers increasing the playback signal to the level required for reproduction at the loudspeaker.

To summarise, in the record position the audio input is fed via the record amplifier to the record head, the playback and a.f. amplifier disconnected. The erase head and record head are brought close to the tape, the tape running past the erase head immediately prior to the record head. On playback the record/playback head is connected to the play-back and a.f. amplifier, the record amplifier, erase head and bias oscillator being disconnected. In this position the erase head is moved away from the tape and carries no 'erase' signal.

Digital Clock

Time measurement in clocks and watches requires an event which occurs repeatedly at regular and constant intervals. In early clocks, and in some modern ones, the 'event' is the movement of a pendulum which swings to and fro at regular intervals determined by the length, size and mass of the pendulum. To maintain this periodic movement a small input of energy is required again at regular intervals, to overcome friction and other losses which would stop the motion. To understand this one has only to consider a child's swing where, unless it is given a 'push' on each occasion it reaches its full movement in one direction, it eventually stops. In mechanical clocks and watches the 'push' or energy input is provided by the force exerted by a taut spring unwinding under controlled conditions. In electric clocks and watches the energy is provided, again at regular and controlled intervals, by the electrical energy source. Display of time may be achieved using 'hands' pointing to figures which are distributed around the clock face or using a system of figures which change as the time changes. The first kind of display is called 'analogue' because the distance around the clock face is used to represent time (analogue meaning one quantity used to represent another), the second kind of display being called digital because the actual figures or digits giving the time in hours, minutes and, sometimes, seconds are continually indicated at the clock or watch face.

In electronic digital clock systems the regularly occurring event used as the reference by which the passage of time is measured is a series of electrical pulses produced by a crystal controlled oscillator. The output from such an oscillator is regular and extremely accurately maintained, which can result, in the case of electronic clocks, in an accuracy to within one ten-thousandth of a second over a period of months. The crystal is made of a material, usually quartz, which exhibits an effect called the *piezoelectric* effect. 'Piezo' means 'pressure' and the piezoelectric effect is the production of a voltage when pressure is applied to the crystal. Apart from clocks or other electronic systems, the effect is now widely used in so-called electronic gas lighters, a spark being obtained from the crystal and used to ignite the gas. The piezoelectric effect is reversible in that the application of a voltage causes a mechanical movement. The crystal itself has a natural frequency of oscillation, determined by the size of the crystal and how it is cut, and once set into motion it expands and contracts at this frequency, an electrical energy source being used to apply a small voltage periodically and by the use of the piezoelectric effect to maintain the mechanical vibration. The frequency of vibration is virtually constant and it is this factor which ensures the high level of accuracy of the clock or watch in which the crystal is used. If the crystal is now combined into an electronic oscillator circuit we can obtain a series of regular pulses which may be used to record the passage of time. The natural frequency of oscillation and therefore the number of pulses produced per second is far higher than one (a common figure is 32.768 kHz) and so a divider or series of dividers is necessary to give the seconds, minutes and hours count. A divider circuit is one which emits one pulse after a number of input pulses.

An arrangement of units for an electronic digital clock is shown in figure 10.15. The crystal controlled oscillator is connected via a series of dividers to give 1 pulse per second, 1 pulse per minute and 1 pulse per hour respectively. Each of these signals is fed via a decoder to the display unit. The decoder is a unit which applies the necessary voltages to the segments of the display device in order to make up the number or digit to be displayed (display devices are described in more detail in chapter 11) according to the number of the pulse received. If all the displays are set to zero after one second the seconds display reads 1, after two seconds it reads 2 and so on up to 59. At this point the seconds decoder changes the display to zero again and the pulse from the minutes divider circuit causes the minutes display to read 1. After 11 hours, 59 minutes, 59 seconds all displays are set to zero for a 12 hour display (or after 23 hours, 59 minutes, 59 seconds for a 24 hour display). For a 12 hour display an additional indicator is sometimes used to distinguish a.m. from p.m. To set the clock the seconds pulses or a set of pulses greater than one per second may be used to speed up the display from zero : zero : zero to the actual time. The display is then held until synchronisation with a time reference can be made. The 'speaking clock' provided by the Post Office is a useful reference for most purposes.

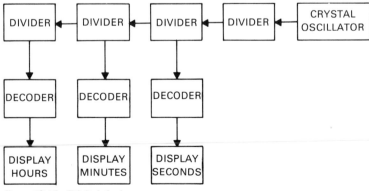

figure 10.15   Digital clock

11 Components

Chapter 10 was concerned with the various types of electronic subunits that make up a system. These subunits are themselves made up of a number of *electronic components* connected together to form electrical current paths called a *circuit*. A diagram showing how the components are connected is called a *wiring diagram* if it shows the exact location of the components and the wires connecting them, and a *schematic diagram* if it shows how the components are connected but does not show the physical location of the components within a subunit. To enable a drawing of this type to be made, symbols are used for components; and, in the interests of general comprehension, these symbols are laid down by the British Standards Institution (BS 3939: Graphical Symbols for Electric Power, Telecommunication and Electronics Diagrams).

Types of Component

Electronic components may be divided into two types: active components and passive components. Active components include electronic *valves, transistors, diodes* and other devices, most of which are capable of amplifying electronic signals. Passive components include *resistors, inductors* and *capacitors* and are not capable of amplifying signals without the aid of an appropriate active device.

Passive Components

The flow of electric current through a material for a particular value of applied voltage depends upon the structure of the material. In some structures a very high voltage is required to set up even a small current; these materials are called *insulators* and are used whenever electric current flow is to be restricted.

Conductors are materials in which it is fairly easy to establish current, only low voltages being necessary. If the voltage applied to a component is divided by the current through that component the ratio is called *resistance* and it is measured in volts/ampere or *ohms*, for which the symbol is Ω. (See chapter 14 for multiple and sub-multiple units.) A passive component especially constructed to have resistance is called a *resistor*, and types include *carbon composition, carbon film, metal oxide* and *wirewound* (the name describing the type of material used). Resistors may be fixed or variable, and the construction details of certain types accompanied by their symbols are shown in figure 11.1 Large fixed

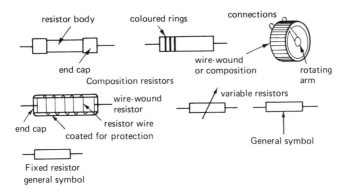

figure 11.1 Fixed and variable resistors

resistors have their resistance value written on the outside in figures, but the value is usually indicated by coloured rings (see appendix 2 for the colour code used) if they are of the smaller fixed resistance type. An important characteristic to be carefully watched is the *power rating* of a resistor, as components operated at greater power levels than their rating will burn out.

The second passive component to be considered is the *capacitor*. A capacitor is a device that comprises one or more sets of conductive plates separated by a good insulator. Capacitors are specially made to exhibit *capacitance*, which is the ability to store electric charge. A capacitor in a d.c. circuit will charge up, and when fully charged current flow stops. Thus, in normal use a good capacitor has a very high resistance to d.c. and may be used to block d.c. signals. The opposition to alternating current, however, is not as high since the conductive plates alternately charge and discharge. Opposition to a.c. is called *capacitive reactance*, symbol X_c, and it is measured in ohms. Capacitive reactance is given by $X_c = 1/2 \, \pi f C$ where f is the frequency (hertz) and C is capacitance, measured in *farads*. (See chapter 14 for multiple and sub-multiple units.) One farad (symbol

F) means that one coulomb of charge is stored for every volt applied. It follows from the formula just given that a capacitor offers an opposition which *falls* as the signal frequency *rises*; and, consequently, a capacitor may be used to block d.c. but still allow a.c. signals to pass. There are various types of capacitor, including *paper, ceramic, mica* and *electrolytic*

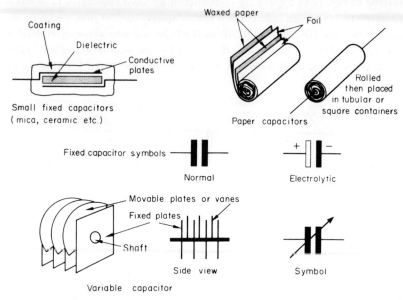

figure 11.2   Fixed and variable capacitors

(the names describing the insulator used between the plates). Variable capacitors also are available, in most of which moving plates are used to alter the capacitance. For typical construction of fixed and variable capacitors together with their appropriate symbols see figure 11.2. Care must be taken when using capacitors to ensure that their *rated working voltage* (shown on the case) is not exceeded and that the poles of electrolytic capacitors are correctly connected to the d.c. supply.

An *inductor* is a passive component specially designed to have *inductance*, which is a measure of opposition to *changing* current. An inductor connected in a d.c. circuit will slow down the *rate of rise* of current when the switch is closed. Once steady conditions are established the current settles at a value determined by the applied voltage and the inductor resistance. The property of inductance is due to *electromagnetic induction*, which is the setting up of a voltage across a conductor whenever a magnetic field around the conductor is changing. Since any electric current sets up a magnetic field, a *changing* current produces a changing field and thus an induced voltage. This induced voltage tries to stop the change causing it, so that an inductor opposes changing current.

As increasing the magnetic field set up by a current increases the induced voltage and therefore the inductance, the inductance can be increased by wrapping the wire in the form of a coil. The field (and inductance) may be further increased by wrapping the coil round a magnetic material such as iron. Thus, high valued inductors are heavy components made up of many turned coils and iron cores. At high frequencies much energy is lost in an iron core and so iron-dust or air cores are used. The inductance of a coil may be varied by moving the core to weaken or strengthen the field. Typical construction and symbols are shown in figure 11.3. The inductance of any coil is measured in *henrys*, symbol H. (See chapter 14 for multiple and sub-multiple units.) A current changing at the rate of *one ampere per second* in an inductor of *one henry* will set up an opposing induced voltage of *one volt*.

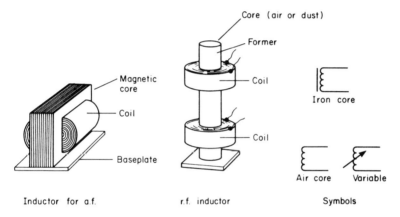

Inductor for a.f.   r.f. inductor   Symbols

figure 11.3   Inductor construction and symbols

In a d.c. circuit an inductor slows down the change of current on switching *on* and again on switching *off*. In the steady state the current is not changing and is affected only by the resistance of the coil wire. However, a.c. is changing all the time, and thus an inductor offers a continual opposition to a.c. This opposition is called *inductive reactance*, its symbol is X_L, and it is measured in ohms. X_L is related to frequency f (hertz) and the inductance L (henrys) by the equation $X_L = 2\pi f L$ and therefore an inductor offers an opposition that *rises* as the signal frequency *rises*. Thus an inductor is used either to oppose the passage of a.c. signals, or else to develop a high voltage between its terminations when an alternating current is passed through it.

Active Devices

Active devices may be divided into two types: gas-filled or vacuum devices and solid-state devices. In the first type, the electronic *valve* (called a *tube* in the USA), the flow of electric charge carriers takes place in a vacuum or in an inert (inactive) gas at low pressure. In the second type current flow occurs within a specially prepared solid called a *semiconductor*. All active devices have two or more points to which connections are made and these are called *electrodes*. Two electrode devices are called *diodes*, three electrode devices are called *triodes*. In addition there is a tetrode, pentode, hexode and heptode with four, five, six and seven electrodes, respectively. It should be noted that these names are normally reserved for thermionic or gas-filled devices, multielectrode solid-state devices are given special names, usually, although not always, grouped under the general name *transistor*.

Diodes

A diode has two electrodes, called the *anode* and *cathode*. The resistance offered to the flow of electrons through a diode is high from anode to cathode and low from cathode to anode. The diode is thus a *unidirectional* device allowing a much higher current in one direction than in the other. Valve diodes fall into two categories, hot cathode (thermionic) and cold cathode. In the thermionic valve the cathode is heated by the passage of a separate electric current (not the valve anode–cathode current) through a *heater* or *filament* and electrons are emitted from the hot cathode. The cathode may be the heater itself or may be a separate electrode wrapped around the heater. If the heater acts as a cathode the valve is called directly heated and if the cathode is separate the valve is called indirectly heated. The construction of and symbols for these valves are shown in figure 11.4. Cold cathode diodes do not have a heater and rely for their current on the movement of free electrons contained within an inert gas. These valves are normally used for voltage stabilising since it is found that under certain conditions the anode–cathode p.d. remains substantially constant over a wide range of currents.

Semiconductor diodes consist of two layers of specially prepared semi-conductor material, called *p*-type and *n*-type respectively. No heaters are required and the unidirectional action of the diode takes place because of internal electric fields set up within the solid when the two types of material are adjacent. As with valve diodes certain types of semi-conductor diodes may be used for voltage stabilising. These are called *zener* diodes. Symbols for semiconductor diodes are shown in figure 11.4.

Other Multielectrode Valves

The triode valve has *three* electrodes. In addition to the anode and cathode a third electrode, called the *control grid*, is interspersed between

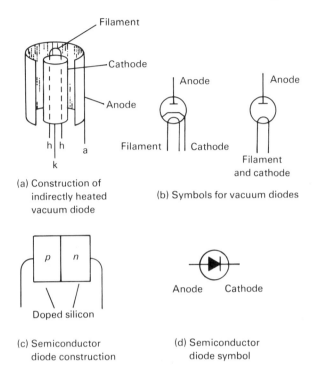

figure 11.4   Diodes

anode and cathode. This grid consists of a fine wire coil or mesh, the electrons travelling from cathode to anode moving between the wires. It is found that the main valve current can be finely controlled by the application of quite small voltages between the control grid and the cathode. The valve may then be used in the process of signal amplification. The voltage/current characteristics of the triode valve may be changed by further grids placed between the control grid and the anode. These grids include the *screen* grid and the *suppressor* grid. A tetrode valve has anode, cathode, control grid and screen grid (four electrodes). A pentode valve has these electrodes and also a suppressor grid (five electrodes). There are also other grid arrangements to make the six electrode, or *hexode*, valve and the seven electrode, or *heptode*, valve (see figure 11.5).

Other Multielectrode Solid-State Devices

A semiconductor diode is made up of two layers of specially prepared semiconductor material. Addition of further layers, suitably arranged, produces other multielectrode devices which can carry out most if not all

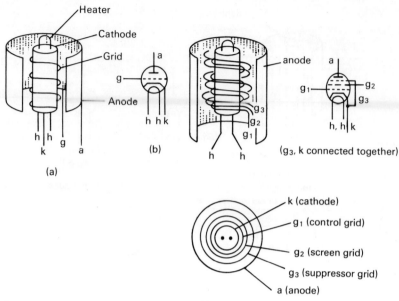

figure 11.5   Valves

The pentode valve

the functions previously reserved for valves. These devices are normally grouped under the name *transistor*.

There are a number of types of transistor, the differences between them being largely one of layer arrangement and thus internal operation. The two main types are *bipolar* transistors and *unipolar* or field effect transistors.

Bipolar transistors consist of three layers of semiconductor material, called the *emitter, base* and *collector* (corresponding in some ways to cathode, control grid and anode of a valve). As indicated earlier there are two types of semiconductor used in electronics, *n*-type and *p*-type. If the bipolar transistor has layers of *n*-type, *p*-type, *n*-type it is called an *npn* transistor and, similarly, if the layers are *p*-type, *n*-type, *p*-type it is called a *pnp* transistor. Construction and symbols are shown in figure 11.6. In both types of bipolar transistor small voltages applied between base and emitter control the current flow between the emitter and collector.

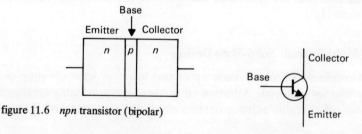

figure 11.6   *npn* transistor (bipolar)

Field effect transistors employ a conductive channel of one type of semiconductor set within a block of the alternative type of semiconductor as shown in figure 11.7. There are again two main types, each of which is further divided depending upon the type of semiconductor used in the channel and on the mode of operation. In all types of field effect transistor there are three electrodes called the *source, gate* and *drain*. In

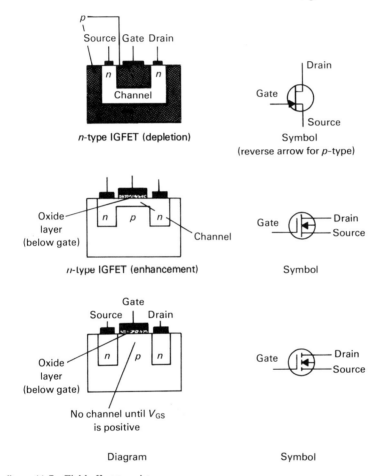

n-type IGFET (depletion)

Symbol
(reverse arrow for p-type)

n-type IGFET (enhancement)

Symbol

Diagram

Symbol

figure 11.7   Field effect transistors

the junction gate field effect transistor or JUGFET there is a permanent channel, as shown in figure 11.7, and if it is in n-type material the transistor is called an n-type JUGFET. Similarly a p-type JUGFET has a permanent channel in p-type material.

The other type of field effect transistor is the insulated (or isolated) gate FET, or IGFET. In this type the gate is not directly connected to the channel but is separated from it by an insulating layer. This layer consists of oxide which gives rise to the alternative name for this type of transistor—metal–oxide–semiconductor FET or MOSFET. There are two types of IGFET, the *depletion mode*, in which there is a permanent channel, which may be reduced in width (depleted) by the gate-source voltage, and the *enhancement mode*, in which there is no permanent channel, the channel being created (enhanced) by the gate-source voltage. As before there are *p*-types and *n*-types of all these varieties depending upon the nature of the channel and the surrounding semiconductor (see figure 11.7).

In all types of field effect transistor the current flow between source and drain is controlled by small voltages applied between gate and source (in much the same way as a valve cathode–anode current is controlled by the control grid–cathode voltage) and the devices may be used for low voltage (or current) control of higher voltages and currents and for amplification.

Thyratrons and Thyristors

These are electronic components that have the common characteristic of being switched into conduction by a control grid or gate, which then loses control. Switching off either device may then be accomplished only by lowering the anode–cathode voltage. A development of the thyristor is the *triac* (which has no direct valve equivalent), which has controlled

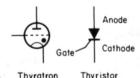

figure 11.8   Thyratron   Thyristor

conduction in both directions. All these devices are used in high power switching circuits (using low power pulses) and may be used to control power without the waste of energy which takes place when variable resistors are used (see figure 11.8).

Photoelectric Devices

In thermionic valves energy is required to free electrons within or from a material, those receiving sufficient energy being then able to leave the cathode altogether. Photoelectric devices use light as the energy source. They fall naturally into three cateogories: *photo-emissive*, *photo-conductive* and *photovoltaic*.

Photo-emissive devices are those in which energy from incident light radiation causes electron emission. The amount of the emission depends not only on the level of incident light but also on the materials used. Common materials used are: caesium–antimony, which responds best to light at the violet end of the light spectrum; calcium sulphide, which has maximum response to red light; and caesium–silver, which is most responsive to infrared radiation. The reason the colour is important is that the energy carried by any electromagnetic radiation is dependent to some extent on frequency, and therefore on the wavelength of the radiation. Different materials require a different amount of energy for the release of electrons (called the *work function*) and thus respond more to certain

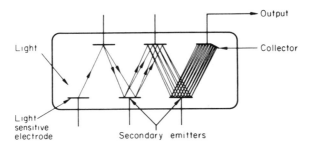

figure 11.9   Photomultiplier

wavelengths than to others. The current flow in photo-emissive cells may be increased by using a gas filled device, in which initial electrons cause ionisation of the gas and subsequent release of many more electrons; or by using a *photomultiplier* technique, in which primary electrons released by light are directed on to a target that reflects the beam and adds to it by *secondary emission* (due to the absorption of energy by the secondary emissive surface). Such a process may be continued to give a very large output current (see figure 11.9).

Photoconductive cells contain a material having internal characteristics such that electrons are not fully released but are nevertheless freed from their molecular bonds, to become available for charge carrying (that is, to form current) and thus the conductivity of the material increases. This effect is used in light meters in photography and is also the basis of the photodiode and photo-transistor. The saturation (reverse) current of a PN diode is sometimes light sensitive, so that if the diode is mounted in a transparent container and reverse biased it may be used as a light controlled switch. The photo-transistor employs the same principle and in addition amplifies the light dependent current. If this effect is not required solid-state diodes and transistors are mounted in opaque containers.

Certain semiconductor materials (for example, silicon, boron and selenium) when arranged correctly set up an e.m.f. on exposure to light radiation. This effect is called the *photovoltaic* effect and is used in light meters that do not require any additional power source.

The photoelectric devices considered so far receive energy from light transmission and produce an electrical effect, for example conduction or emission. There are other devices which work in an opposite sense, receiving energy from an electrical source and producing light. Two such devices are the *light emitting diode* and the *liquid crystal display*.

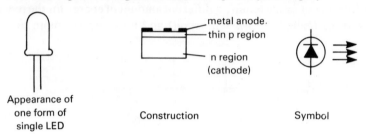

figure 11.10   LEDs (light emitting diodes)

The *light emitting diode* (LED) is a two-layer semiconductor device which, when a voltage is applied to it in the correct sense, emits light. The colour of the emitted light depends upon the material used within the diode and currently red, green and yellow are the available colours (see figure 11.10). The LED principle is used in *seven segment* displays used for producing figures and/or letters in calculators and other indicating equipment (see figure 11.11).

Forming numbers with 7-segment LED

seven segment LED

figure 11.11

The *liquid crystal display* (LCD), used similarly in indicating instruments, works on a slightly different principle in that it does not generate light as the LED does but uses incident light from the surroundings. The LCD when energised either reflects the light or allows it to pass through, giving a different appearance in each case to the observer. The LCD requires a much lower energy input than the LED (since it does not produce light) but cannot be seen if the surroundings are dark. In this case artificial lighting is provided (for example the 'back-light' in an LCD wristwatch) (see figure 11.12).

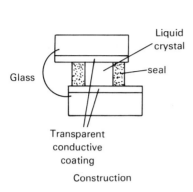

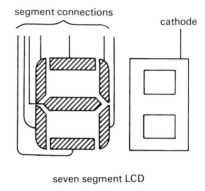

figure 11.12   Liquid crystal displays (LCDs)

Integrated Circuits

Until a few years ago all electronic circuits were made up of individual (or *discrete*) components connected together with wires. These connections were either separate conductive leads, or else were copper strips arranged on a printed-circuit board (that is, a copper-coated insulating board, such as bakelite, from which all surplus copper has been etched leaving only the required circuit pattern). With the rapid development of solid-state devices during the early 1960s, however, it became possible to manufacture more and more components in a single piece (or *chip*) of semiconductor material. At the same time development proceeded on smaller and smaller discrete components to be used with the new varieties of *film circuits* then becoming available. It is now possible to think in terms of 10 000 components per cubic centimetre, using one or more of the various *integrated circuit* techniques. A single integrated circuit can already contain almost a complete radio receiver, four or five a complete television receiver. Computers, which once required several cubicles two or more metres high are now contained in a single desk unit.

Integrated circuits may be divided into two kinds, *film* circuits and *monolithic* circuits. A film circuit is made by printing patterns of conductive (or resistive) material on to a ceramic base, called a *substrate* and firing the mixture in suitable ovens to a hard finish. Using this process it is possible to 'print' resistors, capacitors and to some extent inductors and their connecting leads in one manufacturing process. Microminiature active components or monolithic chips, as described below, may then be added to the finished film circuit using *compression bonding* (exerting controlled pressure at an appropriate surrounding temperature) or some similar technique.

Monolithic circuits are composed of a single chip of semiconductor containing diodes, transistors and passive components (except most inductors) which are manufactured together in the same process. It is now accepted throughout the electronics industry that to take a piece of *p*-type material, dope a layer of controlled thickness of *n*-type material, then repeat the process with *p*-type material in order to make one single *pnp planar* transistor (as shown in figure 11.13) is a time-wasting process when it is just as convenient to make several hundred in the same chip. In fact, this process is now used to make discrete planar transistors, the chip being divided by diamond cutters after manufacture. *pn* diodes, *pnp* and *npn*

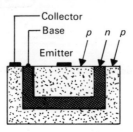

figure 11.13   *pnp* planar transistor

bipolar transistors and FETs may all be made using this planar process, the impurity required to produce *n*-type or *p*-type properties being introduced by a *diffusion process* (involving passing a hot gas containing the additive over the surface for a controlled period of time). Resistors in a monolithic chip are a layer of either *p*- or *n*-type material suitably doped to give the required level of resistivity. Capacitors consist of two conductive layers separated by an appropriate oxide layer to act as dielectric. Spiralling a conductive layer to increase inductance is a possibility but most inductors are added externally, the problem here being the core material. Care should be taken when attempting to understand the equivalent circuit of an integrated circuit supplied by a

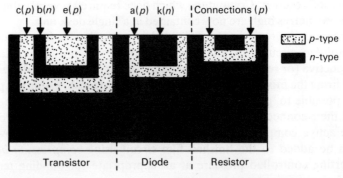

figure 11.14

Integrated circuit components

manufacturer, because it is common practice to use the resistive or capacitive properties of (say) a transistor instead of using an actual resistor or capacitor, the reason being that it is often easier and cheaper to mass-produce hundreds of transistors simultaneously and then print the interconnecting pattern of conductive leads, rather than to halt and rearrange the process in order to manufacture resistors and capacitors. A small part of a typical integrated circuit is shown in figure 11.14.

Symbols used for integrated circuits depend largely on the function of a circuit. There are a number of possible functions including operational amplifiers, comparators, logic gates and associated circuits and so on.

12 Use of hand tools

The number of basic hand tools in electronics servicing is quite small, although there are a number of additional specialist tools available which should be known and which one should be able to use.

The basic tools are shown in figure 12.1. They consist of side cutters, snipe nosed pliers and a set of screwdrivers. These items are essential and purchase of a personal set is recommended. In addition, although this is usually provided by employers, individuals may prefer to purchase their own soldering iron (described later). Small hand tools (as all tools) should be handled with care, cleaned and where necessary oiled regularly and

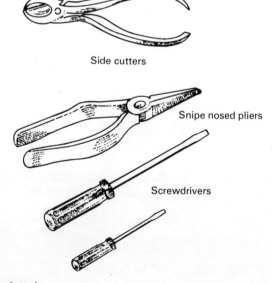

Side cutters

Snipe nosed pliers

Screwdrivers

figure 12.1   Basic tools

not left in a damp atmosphere. The right tool should be used for any particular task and care taken to use the right size where a number of sizes are available. This is the case with screwdrivers where three or four sizes may be carried or may be available for use in a workshop. Trying to turn too large a screw with a small screwdriver ruins the blade, rendering it unusable for its proper function. In the following sections we shall be looking at some basic mechanical tasks carried out in servicing which will use, in addition to the basic tools, specialist tools designed for one particular job. These tasks include making electrical connections and equipment assembly including chassis manufacture.

Wires and Cables

A *wire* is a solid or stranded conductor usually cylindrical in shape and made of copper. It may be bare or covered in an insulating material such as enamel, rubber or plastics, a commonly used plastics being polyvinyl chloride, abbreviated PVC. A *cable* consists of a number of wires alongside one another, the whole being covered in insulation. (It should be noted that the term cable is also often applied to very thick single wires used for carrying high voltages and currents in power engineering applications.) Care must always be taken in selecting wires and cables for any particular use, attention being paid particularly to current and voltage carrying ability and any special requirements in insulation or covering.

Single strand wire is classified according to the standard wire gauge (S.W.G.) which allocates a number according to the wire diameter. S.W.G. tables are normally available in electrical and electronic workshops; an extract from these tables is given in table 12.1. As can be seen the higher the gauge number the smaller the diameter. It follows that the resistance per given length increases as the S.W.G. number rises and the current carrying ability is reduced as is shown in table 12.2. Multi-strand wire is classified according to the number of strands and the diameter of each strand. Thus 7/0.2 indicates seven strands, each of 0.2 mm diameter. A wide range of multistrand wire is available of varying strand numbers and diameter and insulating cover.

S.W.G.	Diameter (mm)	S.W.G.	Diameter (mm)
5	5.38	30	0.3
10	3.25	35	0.21
15	1.83	40	0.12
20	0.9	45	0.07
25	0.5	50	0.025

table 12.1

S.W.G.	Resistance per kilometre (Ω)	Maximum current (A)
10	2.077	13
20	26.26	1
30	221.35	0.12
40	1476.8	0.018

table 12.2

A cable consists of a number of wires, or *cores*, each of which may be single or multistrand and is insulated from its neighbour. The cable is itself covered in an insulator which, as before, may be rubber or plastics. Thus, a typical mains cable for carrying up to 3 A at 230 V–250 V, 50 Hz may consist of three cores, each core being a 16/0.2 wire (16 strands per wire, each of diameter 0.2 mm).

The outer insulator or *sheath* of such a cable could be PVC, rubber or rubber/braid, the braid being a further covering of cellulose or textile.

Special cables used mainly for carrying electronic signals consist of a central single or stranded wire conductor surrounded by polythene insulation, which in turn is surrounded by copper braiding, the whole being encased in PVC. These cables are called *coaxial cables*, the control conductor carrying the signal and the sheath (braiding) being connected to ground or to the common point. There are various physical sizes of

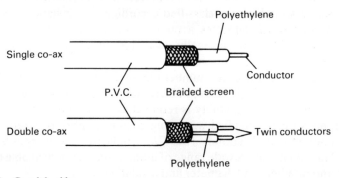

figure 12.2   Coaxial cables

these cables, choice being determined by the effective resistance (or impedance) presented to the signal and the resulting loss in signal strength as it proceeds down the cable. There are also twin core coaxial cables carrying two single or sheathed wire central conductors (see figure 12.2).

Within equipment, wires and cables may be identified using a colour code which refers to numbers. The colour code is the same as for single digits in the marking of resistors and capacitors, namely:

Black	0	Green	5
Brown	1	Blue	6
Red	2	Violet	7
Orange	3	Grey	8
Yellow	4	White	9

and by using this any number (corresponding to the number of the wire or cable), may be made up. To indicate colours marker sleeves may be used (reading from the outside inwards) or the outer insulation itself may be coloured, as for example in 10 core round or ribbon cables. In addition, multicoloured insulation may be used carrying the colours in 'streaks' or lines on the outside of the cable. In this case no particular number is assigned to the cable, the function and colour code being listed in the data sheets relating to the equipment (for example amplifier output to output socket: red/blue).

Wire Stripping

When a wire is used to make a connection all insulation must be completely removed from the end of the wire. The method of removal is determined by the nature of the insulation.

For enamel covered wires one of the following methods should be used:

1. rubbing with emery paper;
2. scraping with a knife;
3. burning off with a methylated spirit burner.

Care must be taken not to remove too much insulation; only that which is necessary to leave a piece of bare wire of sufficient length for the connection should be removed. Removal should be thorough since any enamel left on in parts will reduce the effectiveness of the connection.

For PVC or other plastics-covered wires a special wire stripping tool should be used. Two types are shown in figure 12.3. The first type has V-shaped notches to cut the insulation, the jaws being adjustable by a screw and locknut. When using this stripper the screw should be adjusted so that when the jaws are closed the cutting edges cut the insulation but not the wire. The locknut is then tightened and the wire put into the jaws by an amount sufficient to allow the required length of insulation to be removed. The handles are then squeezed together and the stripping tool rotated through half a turn to cut the insulation. It should then be possible to remove the wire without difficulty, the insulation piece falling away. Different diameters of wire are catered for by the adjuster screw. In the second type of stripping tool the wire is held between jaws and a separate cutting edge removes the insulation. The gripping action takes place first

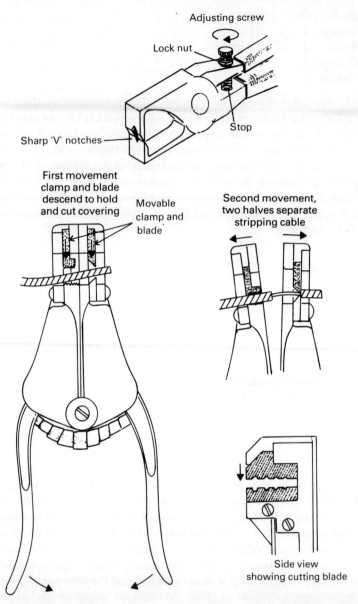

figure 12.3   Wire stripping tools. The lower stripper has an ingenious double action. The first movement of the handles clamps and severs the outer cable sheath. Further movement separates the two halves stripping the sheath from the cable. A variety of blades are available to suit most sizes of cable.

as the handles are pressed and as further pressure is applied the cutting takes place. The cutting edges are circular, different diameter circles being made available to cater for various wire sizes. The correct edge should be chosen for a particular diameter before pressure is applied to the handles.

If the correct tool is not available for plastics-covered wires a penknife or side cutters may be used but much greater care is required since the wire itself is more likely to be damaged with these methods. Partial cuts in the wire may not be immediately obvious but may well show up in the form of bad joints or joints which eventually fail when the bared wire is used to make a connection. Whichever method is employed to remove insulation care should always be taken to remove neither too much nor too little for the connection. Removal of too much will leave bared wire away from the connection, increasing the possibility of touching by other wires or components. Removal of too little leaves insufficient wire for a good connection and, if soldering is used to make a connection, will cause unsightly melting of insulation in the vicinity of the joint. Removal of insulation of coaxial cable is a special case since the copper braiding is

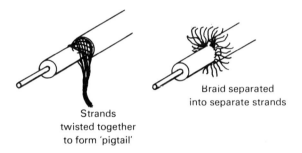

Strands
twisted together
to form 'pigtail'

Braid separated
into separate strands

figure 12.4   Connection with coaxial cable

usually formed into a 'pigtail' suitable for connection at the same time as the centre core is prepared for its connection. There is available a plunging tool which is inserted under the braid and when the plunger is pressed the central core and dielectric is pushed through the braid. If such a tool is not available a penknife or small screwdriver may be used to unpick the braid and separate it into separate fine strands, which may then be twisted together and soldered (see figure 12.4). The dielectric is removed from the central core by one of the methods considered earlier.

Again only sufficient insulation should be removed from the central core to make a good connection and the length of braiding made into the 'pigtail' should be carefully measured.

When preparing multicored cable for connections each core is prepared as described above, the outer overall insulation being removed by the methods described earlier, making a fine cut round the cable or by using side cutters. In any case particular care must be taken not to damage the core insulation at the point where the outer insulation is removed (see figure 12.5).

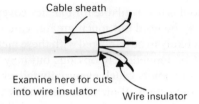

figure 12.5   Multicore cable connection

Electrical Connections

The three main methods of making an electrical connection are soldering, wrapping and crimping, the first method, soldering, being by far the most common. Solder is an alloy of tin and lead with a low melting point. Good quality solder has the proportions 60% tin, 40% lead but a 50/50 mix is commonly used. The solder is melted on to the wire where it forms an atomic bond with the copper, which is then said to be 'tinned'. If two tinned surfaces are put together and heated the solder on them melts and runs together. On cooling a joint is obtained which may be stronger than the original material.

When copper is heated a layer of oxide forms on its surface which would prevent the solder bonding. To prevent this a material called *flux* is used, which cleans the surface and prevents oxidation. Flux may be applied separately but more commonly *cored solder* is used which carries the flux in two or more parallel cores along the solder length (see figure 12.6a).

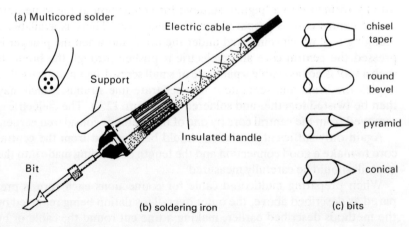

figure 12.6   Soldering

Heating of the wire and the join is usually done by an electric soldering iron, one form of which is shown in figure 12.6b. The part of the iron which comes into contact with the surface is called the *bit*. There are a number of shapes and sizes of bits normally available and with most soldering irons it is possible to change the bit from one kind to another without too much difficulty when the iron is cold (see figure 12.6c). The electric element which heats the iron is rated at a specific voltage and power, the voltage being either mains or some low voltage, the power rating varying according to the size of iron and the task it has to carry out.

Good soldering is an art not acquired without a good deal of experience. The following notes on obtaining a good soldered joint should be carefully studied:

1. Select the correct power rating. Too large an iron may cause damage to the surrounding area, too small an iron may cause poor joints through the solder not melting thoroughly and flowing into the joint.

2. Select the bit to give the best surface contact while having a shape suitable for the best approach to the work. The bit face may be either plated or unplated, iron being used as the plating material. Plated bits last longer, and when finally worn, so that the bit surface is poor, are discarded. Unplated bits quickly become pitted and covered in oxide but may be cleaned and refiled (when cold) to give a good shape and surface. The bit of the iron should be cleaned regularly during soldering work by rubbing with a wire brush (for unplated bits *only*) or special sponge pad (both kinds of bit).

3. To make a good joint, solder must flow evenly over and between the surfaces. This will occur provided that the surfaces are first tinned, are clean and are heated to the right temperature. Sufficient flux is also required for even flow. Hold the soldering iron like a pencil as com-

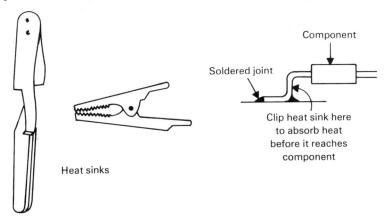

figure 12.7   Heat sinks in soldering

fortably as possible, resting the arm on the bench to improve stability and insert the solder between the job and the iron bit to maintain a flow of flux and solder. Do not keep the iron in contact with the work for longer than is necessary to achieve a smooth flow of solder and once the iron is removed keep the work still until the solder has set. Apply just sufficient solder so that the surfaces are soldered along their full length of contact and the wire shape is still visible through the solder.

4. When soldering delicate components use a *heat sink* on the wire away from the joint, to absorb the heat before it is conducted into the component. If one is not readily available snipe nosed pliers are a useful alternative (see figure 12.7).

When using a soldering iron the following safety points should be borne in mind:

1. The iron should be inspected regularly for damage especially to the lead.

2. The iron should always be handled with care and when not in use placed *in the proper way* in a stand designed for the purpose.

3. Mains irons should be earthed.

4. Solder should not be 'flicked' from the iron, it may harm the user or others around or may fall into the circuit being soldered.

Wrapping wire round a terminal post or other connector is an alternative method of making an electrical connection. The post or connector is usually square or rectangular in cross-section and a special wrapping tool is used to make a firm, tight, wrap of the wire round the post. The

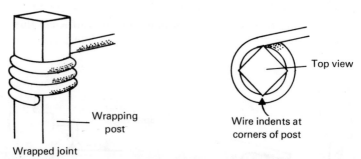

figure 12.8   Wrapping as a means of connection

post edges dig into the wire and lock the joint. Wrapped joints are more commonly found in mass-produced commercial units and without the correct tool it is unwise to attempt a wrapped joint, since it is then difficult to apply sufficient pressure to the wrap (see figure 12.8).

Crimping is another method using conductors pressed firmly together. It is a useful and quick method for obtaining reliable joints, particularly between wires and wire terminations such as connectors for screw terminals. A special tool called a *crimping tool* is required as shown in figure 12.9. A crimped joint is made by inserting the wire, over which the connector is loosely placed, into the jaws of the tool and applying pressure to the handles. Further pressure releases the jaws for the wire to be removed. The jaws are interchangeable thus allowing for a variety of wire and connector diameters. A simpler version of the crimping tool

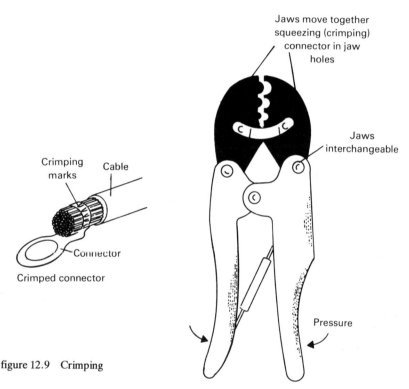

figure 12.9   Crimping

does not have the jaw release facility, the jaws being opened manually by pulling the handles outwards, once the joint is complete. Pneumatic and hydraulic crimping tools are used extensively in mass-production techniques. A good crimped joint can be made only by the correct tool and it is not advisable to use an alternative such as ordinary plier jaws. If a good joint is not obtained with the tool at the first attempt it cannot be remedied except by scrapping the connector and wire (if there is insufficient to prepare a new end) and starting again.

Equipment Assembly

Circuit components are usually mounted on printed circuit boards which are themselves attached to one or more metal frameworks contained within a suitable cabinet. A metal framework holding component boards is called a *chassis*.

Prior to the introduction of printed circuit boards (p.c.b.), components were wired together using individual wires. The introduction of the p.c.b. was a major step in the devising of automatic assembling techniques (mass production) and led to increased reliability of equipment. Basically, a p.c.b. consists of a layer of suitable insulator, such as bakelite, upon which is formed a pattern of copper connections called *tracks* which link circuit components to each other. The pattern is formed using photographic techniques, light sensitive chemicals and etching materials for removing unwanted copper. The board starts its life as a sheet thinly coated on one or both sides with copper. The copper is dipped or sprayed with a material which is light sensitive and etch resistant. A photographic negative (having black areas where the copper is to be removed) is placed

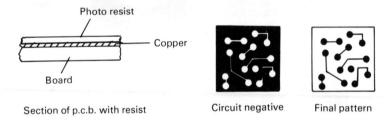

Section of p.c.b. with resist   Circuit negative   Final pattern

figure 12.10   p.c.b. manufacture

on the board and on exposure to ultraviolet light the resist hardens where the light reaches it, that is in the 'white' areas of the negative. This leaves a sheet of resist hardened in a pattern corresponding to the required final pattern of copper. The sheet is now dipped in a material which removes the soft resist and is then covered with an etching liquid that removes copper. Where the resist has hardened, the copper is protected from the etchant and when the board is washed a copper pattern corresponding to the original 'white' parts of the negative is left on the bakelite. An alternative method uses a photograph 'positive' and a nylon screen, etch resistant ink being squeezed on to the board through a pattern made on the screen. The board is then etched in the same way as in the previous method (see figure 12.10).

Printed circuit boards should be handled with care and should not be bent or distorted. A common fault which occurs is the cracking and eventual breaking of the track. Even a 'hairline' crack, often too small to be seen by the naked eye, is sufficient to produce an open circuit and thus a fault condition. Components are mounted on the p.c.b. through small holes, their connecting wires being soldered to the track.

Soldering and desoldering (removal of) components should be done as carefully as possible, the soldering iron not being held on for too long and held so that maximum control over the iron can be effected. Overheating of the tracks due to too hot an iron or one held on too long may cause blistering or lifting of the track. Maximum control is necessary to avoid touching adjacent components or parts of the track not being soldered.

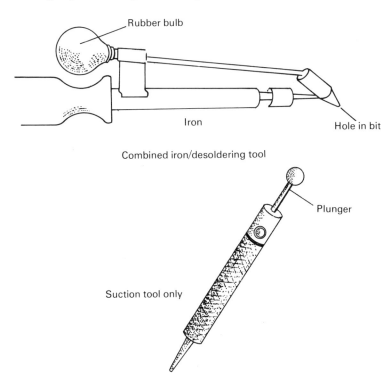

figure 12.11   Desoldering/suction tools

Desoldering components may prove difficult because of the necessary removal of solder around the connection. A useful aid here is a special tool which sucks the molten solder from the board. Two forms are shown in figure 12.11, one which is a combined suction tool and iron, and one which is a suction tool only used separately from the iron. Specially

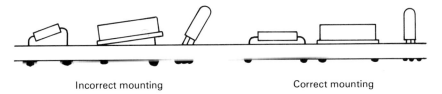

figure 12.12   Mounting components on a p.c.b.

shaped bits are available for desoldering components having multi-connections (for example, integrated circuits) since here all connections must be desoldered at the same time for easy removal of the component. When components are soldered to a p.c.b. they should be mounted as neatly as possible and should lie on or close to the board (see figure 12.12).

Printed circuit boards and other boards carrying components are usually fastened in larger systems to a metal framework called a chassis which is mounted inside the system cabinet. (Small systems, as for example, portable radio receivers use the cabinet itself, which is usually made of preformed plastics, as a chassis.) Metal chassis are made from sheet, which for mass production are stamped out into an appropriate shape, drilled where necessary (to carry p.c.b. racks or large components) and bent into the final shape. It is useful to be able to

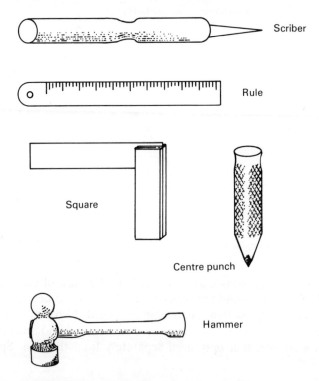

figure 12.13   Tools used in making a chassis

manufacture a chassis by hand for laboratory or test instruments. In this case the sheet metal is marked out using a scriber, metal rule, square, centre punch and hammer (see figure 12.13). Once the external shape is marked, all hole centres are located and punched (with centre punch and hammer) and the shape is then cut out using snips or shears. The work is then firmly clamped into a drill table and holes up to 12 mm are drilled. Larger holes may be cut either by a *trepanning* cutter or a *punch and die set.*

A trepanning cutter is shown in figure 12.14. Non-circular holes are now formed by filing, the work being securely gripped in a vice. The final action in making the chassis is folding to shape, preferably using a

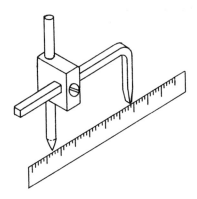

figure 12.14   Trepanning cutter

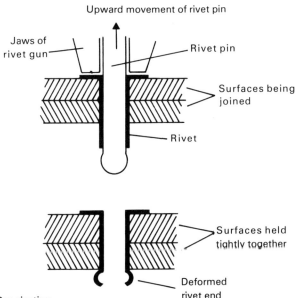

figure 12.15   Pop riveting

Upward movement of rivet pin

Jaws of rivet gun

Rivet pin

Surfaces being joined

Rivet

Surfaces held tightly together

Deformed rivet end

bending machine (although vice jaws may be used for small areas of metal), and securing corners by riveting, brazing or welding. Riveting is most easily achieved using a 'pop rivet gun' as shown in figure 12.15, all that is necessary here being to load the gun, insert the rivet into the hole and press the trigger. Both brazing and welding are similar to soldering in that a metal is melted to run into and join two others. Brazing uses a special brazing rod and gas torch to provide heat, welding uses a special welding rod and may use a gas torch (oxy-acetylene) or electric arc to provide heat. It is not advisable to attempt any brazing or welding (particularly electric arc) without previous experience unless a skilled supervisor/instructor is present and able to assist.

13 Instruments

Efficient servicing of electronic equipment requires a good working knowledge of electrical and electronic instruments, together with the ability to select a suitable instrument for a given application and to use it safely and effectively. In this chapter we shall examine the more commonly available instruments used in servicing and how they are used.

Types of Instrument

Electrical indicating instruments may be broadly divided into two kinds, electromechanical and electronic. Electromechanical instruments are those in which a pointer moves along a calibrated scale (or a scale moves relative to a fixed pointer), the movement being produced by and proportional to the quantity being measured. In electronic instruments there are no mechanical moving parts and the quantity being measured is displayed using electron tubes or semiconductor display devices. The cathode ray oscilloscope (c.r.o.) may be regarded as an electronic instrument since the display device is the cathode ray tube. Both types may be combined to give an instrument using electronics to amplify or otherwise change the input signal (which is produced by the quantity being measured) and the display is made using conventional electromechanical means, that is pointer and scale.

Electromechanical Instruments

In the electromechanical instrument the moving part, pointer or scale, is deflected by a force produced by the quantity being measured. As the moving part is deflected a controlling force is set up, this force being dependent on the movement away from the rest position. When the deflecting force is equal to the controlling force the moving part stops and

191

the quantity being measured can be read off the instrument scale.

When the deflecting and controlling forces are equal the moving part should stop in the new deflected rest position. Without any further form of control it may be found that the moving part does not suddenly stop at the new rest position but because of its momentum is carried past this position. Once past the controlling force is greater than the deflecting force and the moving part eventually stops and starts to move back. Again it may not stop at the point where controlling and deflecting forces are equal (because of momentum) and it moves past. Now the deflecting force exceeds the controlling force and the moving part eventually stops and starts to move again in the opposite direction (see figure 13.1). This oscillation process may continue, making reading of the scale difficult so, in order to prevent this, some form of *damping* is introduced. Ideally the

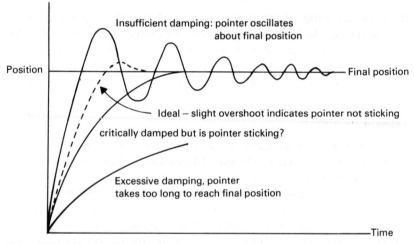

figure 13.1   Effect of damping

moving part should move slightly past the deflected rest position then back to this position. In this way we know that the part has not 'stuck' (as can happen) and can rely reasonably well on the final rest position being determined by the quantity being measured. The most common form of controlling system in electromechanical instruments uses two fine springs wound in opposite directions. As the moving part is deflected one spring tightens and the other unwinds, both springs attempting to return to their rest position and thus exerting the necessary controlling force on the moving part of the instrument. The reason for using oppositely wound springs is that any change in length of spring due to temperature variation is compensated. Damping in electromechanical instruments is usually one of two types, viscous or electromagnetic. Viscous damping systems use a piston or vane attached to the moving part so that during movement

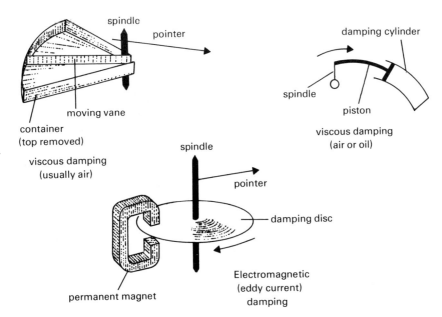

figure 13.2   Methods of damping

either oil or air within a suitable container must be pushed out of the way. This tends to reduce the speed of movement. Electromagnetic systems use the fact that if a metal moves between the poles of a magnet small circulating currents are induced in the metal (electromagnetic induction) and the magnetic fields set up by these currents oppose the pole pieces magnetic field, again slowing down movement. Both types of damping system reduce the oscillation of the moving part about its deflected rest position (see figure 13.2).

There are basically three kinds of deflecting system used in electro-mechanical instruments. These are the permanent magnet-moving coil or d'Arsonval system, the moving iron system and the electrodynamic system.

In the permanent magnet-moving coil (p.m.m.c.) system a coil is suspended so that it can rotate between the poles of a permanent magnet, as shown in figure 13.3. When current is passed through the coil a magnetic field is established by the coil which reacts with the permanent magnet field and the deflecting force is set up, making the coil turn. Usually the controlling force is produced by hairsprings as described earlier and when deflecting and controlling forces are equal the coil stops turning and the moving part, usually a pointer, is at rest.

The deflection from the zero position is directly proportional to the coil current (that is if the coil current is doubled or trebled the deflection is doubled or trebled, respectively) and the scale along which the pointer

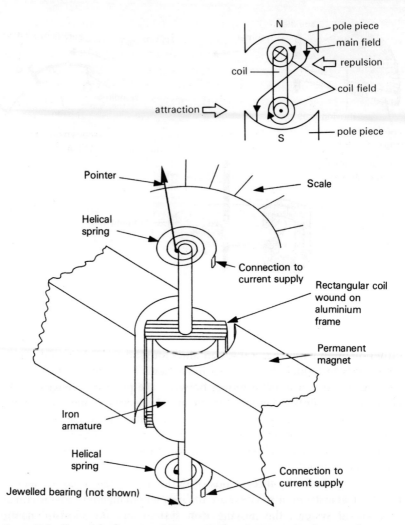

figure 13.3   Essentials of p.m.m.c. instrument using contrawound spring control

moves has its calibration marks equal distances from each other if the quantity being measured is directly proportional to the coil current. Such a scale is called *linear*.

Permanent magnet-moving coil instruments may be used as ammeters or voltmeters to measure quite large ranges by the use of additional resistors suitably connected. The coil voltage and current when the pointer is fully deflected across the scale are usually quite small, being of the order of millivolts and milli or microamperes. For use as an ammeter, therefore, to measure currents in excess of the full scale deflection (f.s.d.) current, a resistor called a *shunt* is connected in parallel with the coil and the excess current passes through the shunt path. For use as a voltmeter to measure voltages greater than the f.s.d. voltage, a resistor called a *multiplier* is connected in series with the coil (see figure 13.4). Multirange instruments able to measure a number of ranges of voltage and current contain a number of shunts and multipliers which are switched into and out of circuit by a range switch (see figure 13.5). The p.m.m.c. instrument

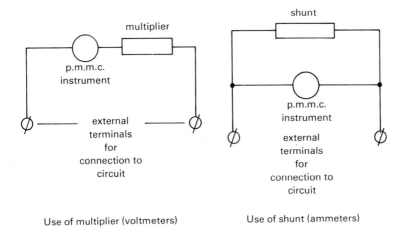

Use of multiplier (voltmeters)   Use of shunt (ammeters)

figure 13.4   Shunts and multipliers

may also be used to measure other variable quantities, for example resistance, provided the scale is marked accordingly and the circuit containing the instrument is arranged to produce a coil current which is changed by the quantity being measured.

For the measurement of currents and voltages or any other quantity which produces a coil current that is directly proportional to the quantity (that is the coil current doubles if the quantity doubles and so on), the scale is linear as described earlier. If the quantity being measured and the coil current are not directly proportional to each other, as for example in the p.m.m.c. ohmmeter, the scale is non-linear and equal increases in quantity on the scale are not marked at equal distances along the scale.

If the coil current is not direct current, that is does not flow in one direction only, the coil will move in a direction which changes as the current direction changes, since, if the current flow through the coil is reversed, the deflecting force is reversed. If the coil current changes its direction too quickly the coil cannot follow it and the pointer and coil vibrate at the zero position. Thus the p.m.m.c. instrument cannot be used to measure alternating currents and voltages unless they are first rectified and made unidirectional. a.c. instruments containing rectifiers may be used to measure voltages and currents and in this case the scale may be

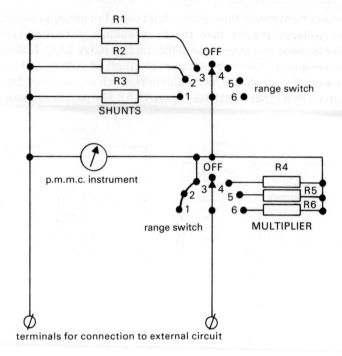

figure 13.5   Multirange p.m.m.c. meter (both halves of range switch mechanically linked so that moving contacts are in same relative position, that is both on 1 or 2 or 3, etc.)

calibrated to read average r.m.s. or peak values of the a.c. quantity being measured. Usually in a.c. instruments the scale reads r.m.s. values.

In the moving iron instrument the deflecting force is again produced electromagnetically, as in the p.m.m.c. instrument. In this case, however, the coil is stationary and the pointer (or scale) is attached to a piece of iron which is free to move within or near the coil (see figure 13.6). There are two types of moving iron instrument, the attraction type and

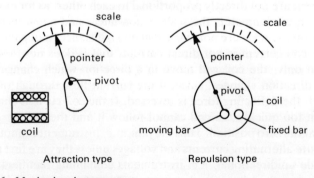

figure 13.6   Moving iron instruments

the repulsion type. In the attraction type the moving piece of iron, shown attached to a pointer in the figure, is pulled towards the coil when current flows in the coil. The moving iron is not magnetised so it is attracted to the coil when current flows regardless of the direction of flow. (Reversing the current flow in the coil reverses the magnetic field—see chapter 4.) The instrument can measure a.c. or d.c. without a.c. having to be rectified as in the p.m.m.c. instrument. The principle of the attraction type moving iron meter is used in door bells and chimes, which may be driven by batteries or bell transformers.

In the repulsion type two pieces of iron are situated inside a coil along its axis, one piece being fixed, the other being movable. When current flows in the coil both pieces of iron are magnetised in the same direction and a force of repulsion is set up, forcing the movable iron piece away from the fixed iron piece. Again, the current direction is unimportant since although a change in direction causes a change in the magnetisation of the iron (that is the end which becomes north seeking when the current is in one direction becomes south seeking if the current is reversed) both iron pieces are magnetised in the same direction and a force of repulsion is set up.

The deflecting force in a moving iron instrument is not dependent directly on the coil current as with the p.m.m.c. instrument, but varies as the *square* of the coil current, that is if the current doubles the deflection is increased fourfold, if the current trebles, the deflection is increased ninefold. This results in a 'square law' scale, which is cramped at one end and 'opens up' at the other. Spring control and damping methods similar to those used in the p.m.m.c. instrument are used in the moving iron instrument. Moving iron instruments are usually used to measure alternating voltages and currents, ranges being changed by the use of instrument transformers. These usually reduce the voltage or current being measured to a level acceptable by the instrument coil, the scale being calibrated to cover the range of the voltage or current being measured. The third kind of electromechanical instrument uses the deflecting force set up between two coils, one of which is stationary, the other mobile, when current is passed through the coils. Each coil sets up its own magnetic field and the reaction between these fields causes a force. By suitably arranging the coils the mobile coil can be arranged to move a pointer or scale. The principal use of this kind of instrument is in power measurement, the instrument then being called a *wattmeter*. The scale is normally a 'square law' scale, as with the moving iron instrument, since the deflecting force is dependent on the product of the currents in the two coils. Control and damping methods are those previously described. If the coils are suitably connected so that both coil fields reverse when the currents reverse the deflection is in the same direction, whether the quantities being measured are alternating or unidirectional. The instrument may thus be used on a.c. or d.c.

The main sources of error common to all electromechanical instruments are friction and reading of the scale. Friction errors are due to wear of the system used for the suspension of the moving part. Friction problems may be reduced by the use of a vertical spindle and horizontal scale. Sometimes the moving part of the instrument is supported by a suspension in place of hairsprings. Such an instrument is called 'pivotless' and is quite robust. It is important to use an electromechanical instrument in the position in which it has been calibrated, usually with scale horizontal. Reading of the scale can cause errors as shown in figure 13.7 and it is essential to read the instrument with the eye directly over the

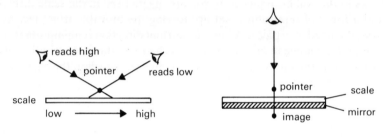

Pointer and image 'lined up'

figure 13.7   Avoidance of parallax

pointer. To assist the reader in knowing when the correct position is attained a mirror is often placed behind the pointer and, once pointer and image are 'lined up', the correct reading position has been reached. This source of reading error is called *parallax*. Errors in p.m.m.c. instruments may occur as the instrument ages and the magnetic field of the permanent magnet weakens. If such an error is suspected, the older meter should be checked against a newer one as a first measure and eventually sent for recalibration if necessary. When used for measuring alternating quantities, errors due to waveform may occur in a p.m.m.c. instrument. As was stated above, in order to measure alternating quantities the current in the coil is made unidirectional so that the deflection will occur in one direction. The coil thus receives a series of unidirectional pulses and the deflection finally produced depends upon the average value of these pulses. The meter is usually calibrated in r.m.s. values. For a sinusoidal waveform the r.m.s. value is equal to the average value × 1.11 (the form factor) and so if, as is most probable, the meter is calibrated using sinusoidal a.c. each reading on the scale will be 1.11 × the average value of the current (or voltage) being read. Consequently, if any other waveform is applied to the instrument an error will occur if the r.m.s. value does not equal 1.11 × the average value. For small alternating quantities rectifiers distort waveforms, due to their own non-linear characteristics, and this reading error is thus far more likely to occur the smaller the alternating quantity being measured.

Electronic Instruments

In this section we shall consider instruments which are wholly electronic both in measurement and display. Electromagnetic instruments discussed above use a means of display consisting of either a pointer moving across a fixed scale or, less frequently, a scale moving past a fixed pointer. In both these methods distance along the scale is marked out in divisions representing the quantity being measured (volts, amperes, ohms etc.). Such a system in which one quantity is used to represent another, in this case distance being used to represent the measured quantity, is called an *analogue* system. In wholly electronic instruments the display is numerical and the quantity being measured is displayed directly using numbers which appear at the instrument face. This type of display is called *digital*, the digits being shown either on semiconductor display devices or, less commonly, on special indicator tubes similar in operation to electronic valves (see chapter 11).

Wholly electronic instruments may be used to measure and indicate values of voltage, current and resistance, as well as other quantities, in much the same way as electromechanical instruments. For most purposes, however, they may be considered to be measuring voltage, additional circuitry being incorporated in the instrument if it is to display quantities other than voltage. In this case the measured voltage will be dependent upon and proportional to the quantity being indicated. The following discussion on electronic voltmeters is therefore equally applicable to wholly electronic instruments used to measure and display quantities other than voltage.

There are a number of different methods of voltage measurement used in electronic voltmeters. Most of them employ some means of comparison of the voltage being measured with a reference voltage by means of a *comparator* subunit. A comparator is a circuit, usually of the integrated variety, which has two inputs and one output, the output voltage level being either 'high' or 'low'. When one of the input voltages is equal to or exceeds the other the output of the comparator changes state and this change may then be used to switch logic circuitry connected to the comparator. Three commonly used methods of electronic voltage measurement using a comparator are discussed in the following paragraphs.

Figure 13.8 shows a system made up of five subunits, a comparator, logic gate, 'clock' oscillator, counter and a digital/analogue converter. The input voltage to be measured is applied to a comparator which compares the measured voltage with a second voltage derived from a

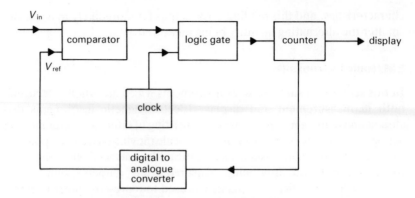

figure 13.8   Electronic voltage measurement using digital/analogue converter.

digital/analogue converter. A digital/analogue converter is a circuit which changes a digital input, consisting of a series of voltage pulses, into an analogue output which is a continuously varying voltage, the variation of the output being determined by the amplitude of successive input pulses and the number of pulses received over a given period. The input to the D/A converter is obtained from a counter, which is controlled by a logic gate. The logic gate output is determined by the comparator output and the output from a square wave oscillator called the 'clock'. When the input voltage is applied to the comparator its output 'opens' the logic gate and allows the clock pulses into the counter. The counter counts the pulses and displays the number on the output display. At the same time the clock pulses are transmitted via the counter to the D/A converter which produces an analogue voltage, V_{ref}, that is then compared with the input voltage. When V_{ref} and V_{in} are equal the comparator output changes state and the logic gate closes. The counter display stops at whatever number has been reached. By choosing the appropriate clock frequency and incorporating scaling circuits (not shown) controlling the display and/or the D/A converter the number at the display may be used to indicate the input voltage directly.

The system shown in figure 13.9 contains similar subunits to the previous system, with, on this occasion, an additional ramp generator. A scaling subunit is also shown. In the previous system of figure 13.8 a voltage derived from the clock is compared with the voltage to be measured until the two are equal, at which point the display reads the value of the voltage being measured. In this system a similar comparison between voltages takes place but the voltage used as reference is derived not from the clock but from a separate ramp generator. A ramp generator is a subunit which produces an output voltage which increases at a constant rate proportional to the input, the waveform (a ramp) giving the subunit its name.

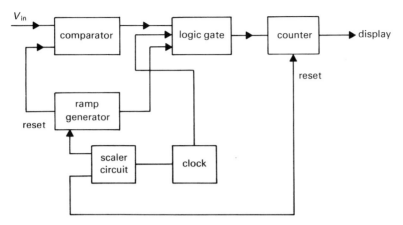

figure 13.9   (a) Electronic voltage measurement using ramp generator: system diagram

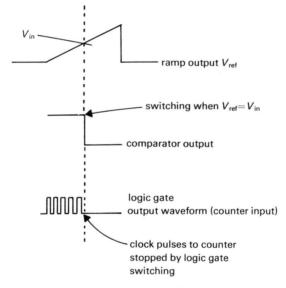

figure 13.9   (b) Waveforms pertaining to figure 13.9(a)

The ramp generator is controlled via a scaling unit by an oscillator acting as a clock, which is also connected to a logic gate supplying a counter and display. The operation of the system is as follows: the output (ramp) of the ramp generator is applied to the comparator together with the input voltage being measured. The comparator output controls the gate, which as long as V_{in}, the voltage being measured, is greater than V_{ref}, the ramp voltage, remains open. The clock output pulses are thus fed from the clock oscillator through the gate to the output display. As soon as the rising ramp voltage, V_{ref}, is equal to the input voltage, V_{in}, the comparator output changes level and closes the gate, 'freezing' the read-out on the display. Again, by arranging appropriate scaling circuits the read-out at this point is arranged to be equal to the input voltage. The ramp is restarted by the oscillator and the counter is reset to zero, the process being repeated between 10 and 100 times per second, thus monitoring any change in the input voltage being measured. The system thus 'samples' the input voltage, the output read-out remaining at the value of V_{in}, because the rate of change of the counter as the ramp rises is too fast to be followed by the eye. The line from the ramp generator to the logic gate is an additional control ensuring that the gate is open only during the period of the ramp. See figure 13.9b for the system waveforms which further clarify the operation of the system.

The final system to be considered is shown in figure 13.10. This system makes use of an input amplifier/attenuator (to increase or reduce the input voltage to the level required for the next unit), an integrator and comparator together with additional units for frequency measurement consisting of a logic gate, 'clock' oscillator and scaler, counter and display. It should be noted that the input amplifier/attenuator shown in this system may also be included in each of the two systems previously considered.

The system of figure 13.10 is a *voltage to frequency* conversion system. In the previous system (of figure 13.9) a steadily increasing ramp voltage from a ramp generator is compared with the input voltage being measured and when the two voltages have the same value the comparator output changes state, controlling the logic gate. In this system a ramp voltage is again used, derived not from a separate generator but from the input voltage being measured via an integrator. The function of an integrator is discussed more fully in chapter 9; for this purpose, we can say that for any particular value of the input voltage being measured, the integrator output is a ramp voltage rising at a rate which is directly proportional to the input voltage. This input-derived ramp is then compared with a separate fixed reference voltage, V_{ref}, by the comparator. When the ramp reaches the same value as V_{ref} the comparator changes state and resets the integrator ready to start the ramp again. The reset is not instantaneous, as shown in figure 13.10b and at the point where the ramp starts again, since the comparator inputs are no longer equal, the comparator output switches again ready for the next ramp. The output of the comparator is thus a series of pulses as shown in the figure (13.10b). The periodic time of the comparator output T is made up of two parts, the period of the ramp rising until it equals V_{ref}, shown as t_1 and the rest period shown as t_2. The length of the ramp time period t_1 is determined by how quickly the ramp rises to equal V_{ref}. Since the rate of rise of the ramp depends upon the value of the input voltage being measured (because of integrator action), the ramp period t_1 and thus the periodic time T

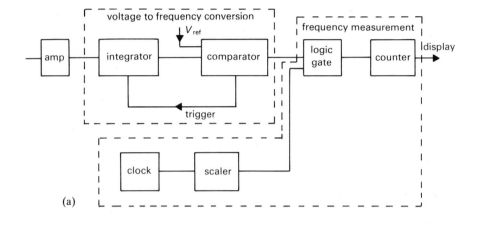

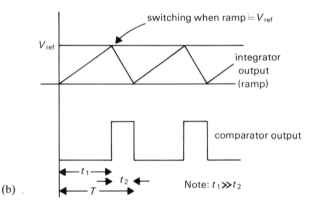

figure 13.10   Electronic voltage measurement by voltage to frequency conversion

depends upon the input voltage being measured.

The comparator output frequency is the reciprocal of the periodic time T and thus it too is determined by the input voltage being measured—hence, the name 'voltage to frequency' converter.

The frequency of the comparator output (and thus the input voltage) is measured by the rest of the system consisting of clock, scaler, gate and counter, the counter output being shown on the display. The clock oscillator opens the gate for a fixed period or number of periods in each second and during each of these periods the counter counts the number of comparator pulses.

The readout thus depends on comparator frequency and thus the input voltage and, as before, by careful choice of scaling circuit components the display can be arranged to read voltage directly.

Electronic digital instruments are extremely precise, easy to read and present an output which is easily stored and processed by a computer, for use in data processing and automatic control systems. Any disadvantage of their greater complexity is offset by the advantage of the greater availability of relatively low cost, reliable integrated circuits which give the instruments greater reliability and reduce the problems of maintenance. There are no moving parts and thus no mechanical 'wear and tear'. Problems due to noise (unwanted signals) and drift (changing voltage levels) have been largely overcome by development of appropriate compensating circuits. As with many other relatively complex electronic systems the cost is either reducing or is being maintained while the instruments are being progressively improved.

Hybrid Instruments

The word 'hybrid' means *mixed*; the third group of instruments to be considered uses a mixture of electronic techniques for processing the input being measured and electromechanical instrument techniques to display the output reading. The readout is thus usually *analogue* and the moving pointer/scale arrangement for display is normally employed.

Most hybrid instruments fall into one of two categories, those which use a rectifier followed by a d.c. amplifier and those which use an a.c. amplifier followed by a rectifier. In both cases a d.c. electromechanical instrument is used to display the output. The possible arrangements are shown in figure 13.11.

Figure 13.11a shows the rectifier/amplifier variety. In this type of instrument both a.c. and d.c. quantities may be measured, the rectifier converting an a.c. input to d.c. and not affecting a d.c. input. After amplification by the d.c. amplifier the output reading is displayed on a d.c. instrument as shown. The second type (fig 13.11b) employs an a.c. amplifier at the input and is therefore used only for measuring a.c. quantities. After amplification the signal is rectified and a d.c. instrument is used for display as before.

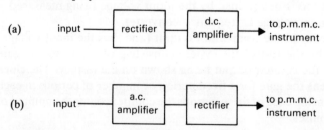

figure 13.11   Hybrid electronic voltmeters

In general hybrid electronic instruments consume negligible power from the circuit in which they are used (and therefore hardly alter the circuit conditions) and may be used over a wide range of frequencies from 10 Hz to 10 MHz. The valve or transistor voltmeter is probably the most widely used instrument for making radio frequency measurements under laboratory or workshop conditions.

Special Instruments

There are a number of instruments frequently used in electronics servicing, both electronic and electromechanical, which have a special purpose. These include signal generators, continuity and insulation testers and the cathode ray oscilloscope. The latter is so widely used and its importance is so great that a separate section devoted to it follows.

Signal generators are used to produce electronic signals for testing and measurements in electronic systems. Signal generators are categorised by the type of signal they produce and the range of frequencies over which they operate. The heart of the generator is an electronic oscillator able to produce a given waveshape and frequency as required. The two most common waveshapes required are the square wave and the sine wave and most commercial instruments are able to produce both, the basic generation being of a sinusoidal wave which may then be processed by additional circuits into a square waveform as needed. Audio frequency signal generators produce an output ranging from 1 Hz to 25 kHz (the normal audio range being up to approximately 20 kHz), radio frequency generators commonly produce output frequencies between 100 kHz and 100 MHz. In addition there are a number of *digital signal generators*, specialised developments of the square wave generators considered earlier, in which it is possible to vary the mark to space ratio of the square wave, and *function generators* to produce any given waveform. Certain of the radio frequency signal generators also have a facility which produces a modulated radio frequency output, the modulating frequency being usually of the order of 1 kHz. A modulated signal is especially useful in testing sections of radio and television receivers.

The permanent magnet moving coil instrument may be easily adapted to measure resistance by the use of a separate d.c. source, such as a battery (contained within the instrument), which passes a current through the resistance being measured. Since the value of this resistance will determine the current and the current determines the meter deflection, the instrument may be calibrated directly in ohms. Such an instrument may also be used to test continuity in a circuit, that is that there is no break in the current path, without actually measuring resistance since, if there is a break no reading will be obtained.

A special form of resistance and continuity tester using a much higher applied voltage obtained from the instrument is the 'Megger' tester. ('Megger' is a trade mark of Evershed and Vignoles Ltd.) In this instrument the pointer is attached to two coils which are rigidly fixed at an angle to one another, the whole arrangement being able to rotate between the poles of a bar magnet.

The supply voltage obtained from the instrument is derived from a hand driven generator and may be of the order of several hundreds of volts. When the instrument output is connected to the circuit being examined the current flow (if any) in the circuit flows through the current coil, the voltage coil also being fed by the generator voltage. The reaction between the magnetic fields produced by the bar magnet and the voltage and current coils causes the coil system and the pointer to deflect. The instrument scale is calibrated to read between zero and infinite resistance (open circuit). Megger type instruments may be used between points in a circuit and ground (the chassis, cabinets or other hardware containing the system should be at ground potential, that is zero volts) and thus the insulation resistance may be tested at a voltage higher than that encountered under normal use.

The Cathode Ray Oscilloscope

Undoubtedly the most versatile of all instruments, the cathode ray oscilloscope is not only able to measure voltage, current, frequency and so on, but a visible picture of the waveform being examined is presented at the same time. The heart of the c.r.o. is the cathode ray tube.

The cathode ray tube has an evacuated glass envelope, suitably shaped, containing a means of producing a flow of free electrons, of controlling the flow and of focusing and deflecting the resultant beam. The beam is directed onto a tube face, or screen, which is coated with a material which becomes luminescent under the impact of the electrons. If correctly focused under no signal conditions a fine 'dot' of light appears at the screen.

In detail, the essential parts of a cathode ray tube are:

(a) a cathode similar to that in the thermionic valve;

(b) a modulator electrode or grid which, by means of a suitable electrical potential, control the beam intensity and thus the brightness of the trace at the screen;

(c) an anode system which is positively biased and accelerates the electrons towards the screen;

(d) a focusing system; this may use electrostatic focusing, in which electric fields control the beam width, or electromagnetic focusing, in which magnetic fields produced by suitable coils, control the beam width;

(e) a deflecting system where again electrostatic or electromagnetic means of deflection may be employed;

(f) a screen coated with fluorescent material.

The parts (a), (b) and (c) are collectively called the 'electron gun'. Whether electrostatic or electromagnetic focusing or deflection is employed is determined by the intended use of the tube. All-electrostatic and all-electromagnetic tubes as well as hybrid varieties are available.

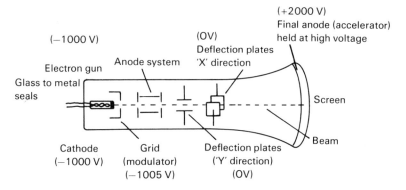

figure 13.12   Cathode ray oscilloscope

Tubes used in television sets or radar displays, for example, use electromagnetic deflection since a wider deflection for shorter tube neck is made available with this method. The cathode ray oscilloscope normally employs all-electrostatic tubes, and a cross-section of a typical tube is shown in figure 13.12.

For the tube shown, electrons are emitted by the heated cathode and are drawn via the three-anode structure (which is positive with respect to the cathode although negative with respect to earth) to the screen. The modulator is negative with respect to the cathode, and by adjustment of its potential the beam intensity and thus brilliance is controlled. The electric fields set up between the three anodes focus the beam as shown in figure 13.13, and the fields and thus focal point are adjusted by altering the potential of the middle anode relative to the other two. There are two

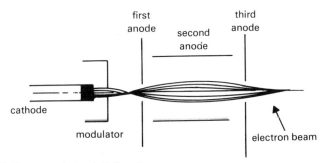

figure 13.13   c.r.o. electrostatic focusing

sets of deflection plates, the X-plates, which are used to deflect the spot horizontally, and the Y-plates, which are used to deflect the spot vertically. The electron beam is attracted towards the more positive of the two plates in each set.

For maximum sensitivity to the deflecting voltage at each set of plates the electrons should not be travelling too fast. On the other hand, the brightness obtained at the screen depends upon the energy and thus the speed of the electrons just before impact. Accordingly, the main acceleration takes place after the beam has passed through the deflecting system and this is achieved by the post deflection accelerating anode shown. This is, in fact, a ring of conductive material around the tube circumference near the screen. The envelope inner surface from screen to third anode is coated with colloidal graphite which effectively acts as a return path from the screen for the electrons. Without this the screen would become negatively charged and tend to repel further electron flow.

The arrangement of the control circuit ensures that the X- and Y-plates are at zero potential with respect to earth whilst still maintaining a high p.d. across the tube from cathode to screen. This ensures safety in operation for the user who will be connecting external circuits to the deflection system.

Use of Instruments

Before making use of any instrument, whatever its purpose, the user should ensure, particularly in the case of multipurpose or multirange instruments, that the function of each switch or other control is known. If in doubt assistance should be sought from the instrument manual or from qualified personnel. Indiscriminate use or experimentation with an unknown instrument can, at best, cause damage to the instrument itself and at worst, risk the life or safety of the user.

To measure voltage, the voltmeter should be connected *across* the voltage being measured, that is in parallel with it. If a multipurpose instrument is being used, it must be connected in the voltmeter mode by setting the appropriate switches *before* connection to the circuit where the voltage is being measured. If the range of the voltage being measured is known (not *thought* to be known) the voltage range may be set prior to connection but in any event it is always advisable to set the range to the highest possible and work down through the ranges until a readable deflection or output display is obtained. *This is essential if the voltage range is unknown.*

An important consideration when selecting a voltmeter for use in a circuit is the effect that insertion of the meter has on the circuit. Ideally the voltages and currents within the circuit should remain the same, otherwise a false reading is obtained. Any voltmeter which draws current from the circuit into which it is inserted must change circuit conditions; the important point to consider is by how much. The more current drawn by the meter the worse will be the effect on the circuit and since the current drawn is dependent on the effective meter resistance, the higher this is the better is the meter. Effective meter resistance is indicated by what is called the *sensitivity* or *ohms per volt* rating of the meter. To obtain the effective resistance this rating should be multiplied by the maximum range voltage (of the range being used). For example, a meter having a sensitivity of 20 kΩ/V on the 10 V range has an effective resistance of 20 $\times$ 10 kΩ, that is 200 kΩ. Clearly the higher the sensitivity of the voltmeter the better is the instrument from the point of view of not upsetting circuit conditions on connection. From this point of view electronic voltmeters (wholly or hybrid) are to be preferred to the electromechanical variety.

As was pointed out earlier, when measuring alternating voltages waveform is an important factor. If the instrument has been calibrated using a sinusoidal a.c. waveform (as is usually the case) any deviation from this waveform will affect the accuracy of reading. The size of the voltage may also be important if it is alternating since, if a rectifier is used, distortion may occur to the voltage waveform. Again, best results for small values of alternating voltages will probably be achieved using electronic voltmeters or the c.r.o.

Ammeters should always be connected *in series* with the circuit or component carrying the current to be measured. Again maximum range should be used and for multipurpose instruments range switches must be set *before* connection. A common 'accident' is the connection of a multipurpose instrument left in the ammeter mode to serve as a voltmeter. The voltage being measured then drives current through the ammeter, and because the meter normally has low resistance the current may be high enough to damage the instrument. Multipurpose instruments should always be left in the no meter position, if there is one, or otherwise in the voltmeter mode. Ammeters do not have a 'sensitivity factor' as do voltmeters but again ideally they should not upset circuit conditions and so should have as low a resistance as possible. In certain low current circuits use of conventional ammeters, even good quality ones, upsets circuit conditions and in these cases current may be determined by measuring voltage across part of the circuit having a known resistance and then using Ohm's Law.

Resistance may be measured using one or other form of ohmmeter, probably the most accurate being those having a digital readout. Before using an ohmmeter it should first be compensated for lead resistance by connecting the leads together and adjusting the 'set zero' control. If this cannot set zero the battery should be changed.

Multipurpose instruments should not be left in the ohmmeter mode for two reasons, the first being the possibility of the next user not checking the mode before use, the second being that, if the leads touch, the ohmmeter will read continually and waste the battery.

'Megger' type instruments should be handled with particular care since they generate a relatively high voltage with consequent potential hazard to safety. Again, these instruments should be left in the 'off' or 'neutral' position when not in use and when in use hands must be removed from the circuit or adjacent chassis or hardware in case of electric shock.

The c.r.o. is one of the most versatile of all measuring instruments. It may be used to measure voltage, current (by voltage measurement), resistance (by voltage and current measurement) and frequency among other quantities and has the additional advantage of waveform display. Before attempting to use a c.r.o. it is essential that the user becomes familiar with the purpose of each control, particularly the following

mains on/off switch
brightness (or intensity)
focus
Y-amplifier gain
X-amplifier gain (if applicable)
timebase frequency
X and Y shift

and the location of X, Y and, if available, Z inputs should be known.

On switching on the c.r.o. if the trace (spot or line) does not eventually appear the timebase frequency control should be switched 'off' and the X and Y shift controls adjusted to mid position. Adjustment of brightness and focus controls should then cause a spot to appear in midscreen. Switching the timebase control should now cause the spot to move horizontally, usually relatively slowly in the low frequency or 'slow' position and faster as the control is adjusted towards the high frequency or 'fast' position. Eventually, a horizontal line should appear on the screen. Signals may now be applied to the Y input, either directly or via the Y amplifier and by adjustment of Y-gain and timebase controls a stationary trace of the required size may be obtained.

A c.r.o. should never be operated with excessive brightness and a stationary trace at high brightness should not be allowed to remain on the screen for long periods. Screen 'burn' may occur if this rule is not followed.

14 Numbering systems, graphs and calculations

In this chapter we shall be concerned with certain topics in mathematics which are used frequently in electronics. Many students regard the study of mathematics as difficult and tedious and, often, even unnecessary. However, certain topics, particularly those listed in the chapter heading, are of particular interest to the student of electronics because they are of direct use in the process of understanding, using and servicing electronic equipment. Indeed, without a basic working knowledge of these topics it is unlikely that much progress can be made in gaining a *working* knowledge of electronic systems.

Although this chapter is arranged as the last it should not be studied after the other chapters but in conjunction with them. Most of the examples given are directly related to topics covered elsewhere in other chapters.

Raising Numbers to a Power

When a number is multiplied by itself once or more than once the process is called *raising to a power*. For example

5 raised to the power 2 is 5 × 5
5 raised to the power 3 is 5 × 5 × 5
5 raised to the power 4 is 5 × 5 × 5 × 5

Notice that the number being raised to the power appears a number of times equal to the power (twice when raising to the power of 2, three times when raising to the power 3 and so on).

Notice also that the number being raised to a power is multiplied by itself a number of times equal to one *less* than the power. In the above example 5 raised to the power 2 is 5 multiplied by itself *once*, 5 raised to the power 3 is 5 multiplied by itself *twice* and so on. Mathematically we indicate 'raised to the power 2' by writing 2 above and to the right of the number being raised, this 5^2. Similarly 5^3 is 5 raised to the power of 3, 5^4 is 5 raised to the power of 4.

Raising to the power of 2 is called 'squaring' so that 5^2 is called 'five squared'. Similarly 5^3 is called 'five cubed'. Special terms are not used for raising to other powers.

Provided that the number being raised, often called the *base*, is the same in any expression containing a number of such terms, powers may be added if the terms are being multiplied, or subtracted if the terms are being divided.

For example

$$6^4 \times 6^2 = 6^{4+2} \text{ that is } 6^6$$
$$6^4 \div 6^2 = 6^{4-2} \text{ that is } 6^2$$

Remember, however, that this rule only applies if the base is the same in each term. The powers in terms having different bases cannot be added to each other or one subtracted from the other. This can be seen by writing out in full the above example

$$6^4 \times 6^2 = 6 \times 6 \times 6 \times 6 \times 6 \times 6$$
$$= 6^6$$

and

$$\frac{6^4}{6^2} = \frac{6 \times 6 \times 6 \times 6}{6 \times 6}$$
$$= 6 \times 6 \text{ by cancelling}$$
$$= 6^2$$

but

$$\frac{6^4}{5^2} = \frac{6 \times 6 \times 6 \times 6}{5 \times 5}$$

and the denominator cannot be cancelled into the numerator.

Raising to the Power Unity and Zero

Raising a number to the power unity gives the number itself since this means *not* multiplying the number by itself. This follows from the following example

$3^3 = 3 \times 3 \times 3$ (multiplied by itself twice)
$3^2 = 3 \times 3$ (multiplied by itself once)
$3^1 = 3$ (*not* multiplied by itself)

On the face of it raising to the power zero appears nonsensical. We must, however, know what it means since in an example such as

$$6^3 \div 6^3$$

we have

$$6^3 \div 6^3 = 6^{3-3}$$

$$= 6^0$$

using the subtraction of powers rule given earlier.

Clearly $6^3 \div 6^3$ is unity since any number or term divided by itself is unity.

Thus, raising to the power zero always yields unity, so that

$$1^0 = 1, 2^0 = 1, 3^0 = 1$$

and so on.

Negative Powers

A number raised to a negative power is equal to the reciprocal of the number raised to the same value positive power. (The reciprocal of a number is unity divided by the number). So that 10^{-3} means $1/10^3$ or $1/1000$ (denoted in units by the prefix 'milli', m, as in mA).

Similarly 10^{-6} means $1/10^6$ or $1/1000\,000$ (denoted in units by the prefix 'micro', μ, as in μV).

This is not intended to be a complete set of rules of manipulation of numbers raised to powers. Only the basic essentials are given to enable the student to better understand the remaining chapters.

Further explanations and examples are given in books devoted to mathematics. For our present purposes we will content ourselves with a restatement of the basic rules.

1. Powers in terms containing the same base may be added together if the terms are being multiplied together.

2. If one term containing a number raised to power is being divided by another term containing the same number raised to a power, then the power in the denominator may be subtracted from the power in the numerator.

3. Any number raised to the power unity is equal to the number.

4. Any number raised to the power zero is equal to unity.

5. A number raised to a negative power means unity divided by the number raised to a positive power of the same value.

Systems of Numbers

By far the most commonly used counting system is the decimal or denary system. This system uses as its basic number the number of digits on our two hands—eight fingers and two thumbs—and it is probable that this is the reason for the development of the system since, under normal circumstances, we carry a basic 'counting instrument' with us at all times. It is important to realise, however, that the decimal system is only one of many possible systems, these being determined by the system basic number. There is for example the binary system, using two as its base, which is increasingly used in computer systems. It is also known that some primitive peoples, for example an Amazon tribe called the Yancos and the Temiar people of West Malaysia, use a tertiary system having three as its basic number.

All numbering systems whatever their basic number can be arranged in the same way and have similar characteristics concerning the number of available digits, their arrangement to signify a number greater than the basic number, and so on. The general rules are considered below with particular reference to the decimal and binary systems.

Number of Digits

The number of digits, or 'figures' available, corresponds to the basic number. The basic number is also called the *base* or *radix* of the system. Thus the decimal system, radix ten, has ten digits $0, 1, 2, 3, 4, 5, 6, 7, 8, 9$; the binary system, radix two, has two digits: 0 and 1. (Similarly a tertiary system, radix three, has digits $0, 1, 2$ and an octal system, radix eight, would have digits $0, 1, 2, 3, 4, 5, 6, 7$.)

Notice that for all systems there is no single figure to represent the radix, thus for the decimal system the radix ten is represented by 10, that is, a number using two of the basic digits 1 and 0. As we shall see shortly the radix is always represented by 10, and in our study of number systems we have to avoid the natural temptation of thinking of 10 as 'ten'. This is only true for the decimal system.

Numbers Larger than the Base—Order of Digits

In any multi-digit number representing a total count greater than one the least significant figure is the one on the extreme right and represents the number of units in the total count. The number situated one place to the left of the least significant figure represents the number of times the radix is contained in the total count. Thus

for the decimal system 10 means 1 ten and 0 units
for the binary system 10 means 1 two and 0 units
for the octal system (radix eight) 10 means 1 eight and 0 units

Similarly

11 in the decimal system means 1 ten and 1 unit

11 in the binary system means 1 two and 1 unit

Clearly binary 11 means decimal three whereas decimal 11 means decimal eleven.

The figure situated two places to the left of the least significant figure represents the number of times the square of the radix is contained in the total count. Thus

for the decimal system 100 means 1 hundred 0 tens and 0 units
for the binary system 100 means 1 four 0 twos and 0 units
for the octal system 100 means 1 sixty-four 0 eights and 0 units

Notice that the words 'hundred', 'ten', 'two', 'sixty-four' apply to the decimal system since it is this system which is best understood. To appreciate this imagine that you have never heard of the decimal system, only of the binary system. The number 11 would not then be understood automatically as 'eleven' but as a number corresponding to decimal 'three'. All systems discussed must be referred to the best-known system —the decimal system—for the value of the count to be appreciated.

A further example of a three-digit number is 111 which

in decimal means 1 hundred 1 ten 1 unit
in binary means 1 four 1 two 1 unit (that is decimal seven)
in octal means 1 sixty-four, 1 eight, 1 unit (that is decimal seventy-three)

Similarly

decimal 101 means 1 hundred and 1
binary 101 means decimal five (1 four, 1 unit)
octal 101 means decimal sixty-five (1 sixty-four, 1 unit)

The figure situated three places to the left of the least significant figure represents the number of times the cube of the radix is repeated in the total count. Thus 1000 decimal means one thousand (ten cubed) no hundreds (ten squared), no tens, no units. 1000 binary means one eight (two cubed), no fours (two squared), no twos, no units.

In general, as one moves one place further to the left from the least significant figure the power of the radix is increased by one and in each case the figure tells us the number of times the radix raised to this power is repeated in the count. For example 10101 decimal means

one ten-thousand (ten raised to the power four)
no thousands (ten raised to the power three)
one hundred (ten raised to the power two)
no tens (ten raised to the power one)
one unit (ten raised to the power nought)

that is ten thousand, one hundred and one whereas 10101 binary means

one sixteen (two raised to the power four)
no eights (two raised to the power three)
one four (two raised to the power two)
no two (two raised to the power one)
one unit (two raised to the power nought)

that is (decimal) twenty-one.

Example 14.1
Write decimal 31 in binary.
   Thirty-one contains 1 sixteen, 1 eight, 1 four, 1 two and 1 unit. Thus decimal 31 = 11111 binary.

Example 14.2
Write binary 1101101 in decimal.
   The extreme figure on the left tells us the number of times the radix (2) raised to the power 6 is repeated in the count, the next figure to the right tells us the number of times the radix raised to the power 5 is repeated in the count. Thus binary 1101101 means

$$1 \times 2^6, 1 \times 2^5, 0 \times 2^4, 1 \times 2^3, 1 \times 2^2, 0 \times 2 \text{ and } 1 \text{ unit}$$

that is

$$2^6 + 2^5 + 2^3 + 2^2 + 1$$

or

$$64 + 32 + 8 + 4 + 1$$

which equals decimal 109.

Binary and Decimal Fractions

As was shown in the preceding section the order of digits, in numbers representing a count of greater than one, determines the number of times that powers of the radix are contained in the total count, the power increasing by one as the figure or digit moves one place to the left from the least significant figure. Exactly the same system may be used to set up a number describing a count of less than one, that is a fraction.
   Some means must be used to indicate the break between those figures representing the number greater than one and those figures representing the fraction. Usually this break is indicated by a dot (or in some European countries a comma) called the *decimal point* for the decimal system, the *binary point* for the binary system, or in general the *radix point*. Thus where 110 for the decimal system means 1 hundred, 1 ten and 0 units, that

is $1 \times (\text{ten})^2 + 1 \times (\text{ten})^1 + 0 \times (\text{ten})^0$ (remember that $(\text{ten})^0 = 1$) then, using a decimal point, the number 110.110 means

$1 \times (\text{ten})^2 + 1 \times (\text{ten})^1 + 0 \times (\text{ten})^0 + 1 \times (\text{ten})^{-1} + 1 \times (\text{ten})^{-2} + 0 \times (\text{ten})^{-3}$

and we see that the first figure to the right after the decimal point tells us the number of times the radix raised to the power -1 is repeated in the count, the second figure to the right tells us the number of times the radix raised to the power -2 is repeated in the count and so on.

Now

$$(\text{ten})^{-1} = \frac{1}{10}$$

$$(\text{ten})^{-2} = \frac{1}{100}$$

so the number 110.110 means

1 hundred, 1 ten, 0 units, 1 tenth, 1 hundredth and 0 thousandths

The last 0 is actually unnecessary.
Similarly the decimal number 2.934 means

2 units, 9 tenths, 3 hundredths and 4 thousandths

The same reasoning applied to the binary system indicates that binary 110.11 means

$1 \times (\text{two})^2 + 1 \times (\text{two})^1 + 1 \times (\text{two})^{-1} + 1 \times (\text{two})^{-2}$

and since

$$
\begin{aligned}
2^2 &= 4 \\
2^1 &= 2 \\
2^0 &= 1 \\
2^{-1} &= 1/2 \\
2^{-2} &= 1/4
\end{aligned}
$$

binary 110.11 means

$4 + 2 + 0 + \frac{1}{2} + \frac{1}{4}$

that is $6\frac{3}{4}$ using a mixed number (whole number plus a proper fraction).

Example 14.3
Convert binary 1.11 to a mixed number.

Binary 1.11 means

$(1 \times 2^0) + (1 \times 2^{-1}) + (1 \times 2^{-2})$

that is

1 + ½ + ¼

which equals 1¾

Example 14.4

Convert decimal 1.11 to a mixed number.
Decimal 1.11 means

$(1 \times ten^0) + (1 \times ten^{-1}) + (1 \times ten^{-2})$

that is

1 + 1/10 + 1/100

which equals 1 11/100 (since 1/10 = 10/100)

Manipulation of Numbers Containing a Decimal Point

Addition

If two or more numbers containing a decimal point are to be added, arrange the numbers one under the other with points in line and proceed as with whole numbers.

Example 14.5

Find the sum of 14.83, 13.7, 11.971

14.830
13.700
11.971
───────
40.501
───────

To help keep the figures in line, noughts have been added to the end of each number. The addition of noughts after the decimal point (but not before) has no effect on the value of the number. As shown, the answer is 40.501

Example 14.6

Add 15, 14.7 and 13.21

15.00
14.70
13.21
──────
42.91
──────

Note that the whole number 15 is written as 15.00

Subtraction

If a number containing a decimal point is to be subtracted from another, arrange the whole numbers one beneath the other with points in line and proceed as with whole numbers.

Example 14.7

Find the difference between 9.76 and 12.53

```
12.53
 9.76
─────
 2.77
─────
```

Example 14.8

Subtract 13.73 from 14

```
14.00
13.73
─────
 0.27
─────
```

Note that as with example 14.6, a whole number—in this case 14—is written as 14.00 to keep decimal points in line. The answer here is a decimal fraction 0.27. Using proper fractions this would be written as 27/100, that is 2/10 + 7/100 (see example 14.4).

Multiplication

Probably the easiest method of multiplication is to convert the figures concerned to whole numbers, multiply in the usual way and determine the position of the decimal point in the answer at the conclusion of the multiplication. First it is necessary to examine the effect of multiplying a number containing a decimal point by ten or powers of ten. Consider

34.7×10

Now 34.7 means 34 and 7/10 as shown earlier. Multiplying 34 by ten yields 340 and 7/10 by ten yields 7

Hence

$$34.7 \times 10 = 340 + 7$$

$$= 347$$

which may be written 347.0 and we see that the decimal point moves one place to the right when multiplying by 10

Thus for example

$$13.93 \times 10 = 139.3$$

$$43.742 \times 10 = 437.42$$

and so on.

Similarly, when multiplying by 10^2, that is 100, the decimal point moves two places to the right. To show this, consider

12.73×100

Now 12.73 can be written 12 and 73/100. Thus

$$100 \times 12\frac{73}{100} = 1200 + 73$$

$$= 1273$$

In general, when multiplying by ten raised to a power move the decimal point a number of places to the right equal to the power; that is one for a multiplier of 10, two for a multiplier of 100 (10^2), three for 1000 (10^3) and so on.

At this point is could be noted that the opposite process takes place when dividing, in that the decimal point is moved to the left by a number of places equal to the power to which ten is raised in the dividing number; that is one place when dividing by 10, two places when dividing by 100, three places when dividing by 1000 and so on.

To return to multiplication, consider the following example.

Example 14.9

Evaluate 1.25×3.5.
First multiply 125×35 in the usual way

```
   125
    35
  ----
   625
   375
  ----
  4375
  ----
```

The problem was 1.25×3.5, the answer 4375 is that of the problem 125×35
Clearly

1.25 has been multiplied by 100
3.5 has been multiplied by 10

The answer 4375 is thus 10×100 times too large. To divide 4375 by 10×100, that is 1000, write 4375 as 4375.0 and move the decimal point three places to the left (three since the dividing number is 1000, that is 10^3). This gives 4.375. Thus

$$1.25 \times 3.5 = 4.375$$

In general, when multiplying two numbers containing decimal points, count the number of figures after the point in each number and add these numbers of figures. Convert the numbers in the problem to whole numbers, multiply out and locate the decimal point in the answer by moving the point to the left, a number of times equal to the total number of figures after the decimal point in the numbers being multiplied. The process is easier to demonstrate than to describe in words. Study the following examples.

Example 14.10

Evaluate 4.2×3.14.

```
    42
   314
   ───
   168
    42
  126
  ─────
 13188
 ─────
```

There is one figure after the decimal point in 4.2 and two figures after the decimal point in 3.14, that is three figures in all.

Write 13188 as 13188.0 and move the point three places to the left giving 13.188. Thus

$$4.2 \times 3.14 = 13.188$$

Example 14.11

Evaluate $4.1 \times 5.6 \times 2.35$.
First two numbers

```
    41
    56
   ───
   246
   205
   ────
  2296
  ────
```

Now multiplying by 235

```
    2296
     235
   ─────
   11480
    6888
    4592
   ──────
  539560
   ──────
```

In 4.1 there is one figure after the point, in 5.6 there is one figure and in 2.35 there are two figures after the point, that is a total of four figures.

Write 539560 as 539560.0 and move the point four places to the left, giving the answer 53.956.

Division

When dividing one number into another and one or both contain a decimal point, the most convenient method is to move the point in both numbers until the dividing number is converted to a whole number. No adjustment of the answer is necessary, since provided the point is moved the same number of places in the same direction in both numbers, the answer remains the same.

For example 12.3/4.1 has the same value as 123/41, since 12.3 has been multiplied by 10 to give 123 and 4.1 has been multiplied by 10 to give 41, that is multiplication of numerator (12.3) and denominator (4.1) by the same figure (10) leaves the answer unchanged.

Example 14.12

Divide 34.644 by 1.2.
This is the same as dividing 346.44 by 12.

```
      28.87
     ───────
  12)346.44
     24
     ───
     106
      96
     ───
     104
      96
     ───
      84
      84
      ──
       0
      ──
```

The answer is 28.87.

Note. The decimal point in the answer lies immediately above the point in the number being divided when this layout of the solution is used.

Example 14.13

Divide 25.74 by 1.5.

Now

$$\frac{25.74}{1.5} = \frac{257.4}{15}$$

```
        17.16
   15)257.4
      15
      ───
      107
      105
      ───
       24
       15
       ──
       90
       90
       ──
       00
       ──
```

The answer is 17.16.

Note. When no further figures are available to be brought down a nought is used. Thus in example 14.13 when 15 is subtracted from 24 leaving the remainder 9 a nought is placed after the 9 to give 90 and the dividing process is continued until there is no remainder. This is quite in order since 257.4 can just as correctly be written 257.400. Sometimes, the process may continue indefinitely with a remainder always present. Consider the following example.

Example 14.14

Evaluate $2.2 \div 0.7$.

This is the same as $22 \div 7$. Using long division (which is not really necessary) we get

$$
\begin{array}{r}
3.14285 \\
\hline
7 \overline{)22.0} \\
21 \\
\hline
10 \\
7 \\
\hline
30 \\
28 \\
\hline
20 \\
14 \\
\hline
60 \\
56 \\
\hline
40 \\
35 \\
\hline
50 \text{ etc.} \\
\hline
\end{array}
$$

In this example the dividing process continues indefinitely and the answer is determined by the degree of accuracy required. This is discussed below.

Addition and Subtraction of Binary Numbers

Addition

There are four basic rules, which are

$$
\begin{aligned}
0 + 0 &= 0 \\
0 + 1 &= 1 \\
1 + 1 &= 0 \text{ carry } 1 \\
1 + 1 + 1 &= 1 \text{ carry } 1
\end{aligned}
$$

The 'carry 1' in the last two lines means that 1 is carried over to the next column on the left when the numbers are arranged one beneath the other. Binary numbers should only be added in pairs of numbers at any one time, otherwise the 'carry' requirement may become confusing, that is, it may be necessary to carry 1 to a column further away than one place to the left.

Example 14.15

Add the binary numbers 101101, 111010, 1011 and check the answer by decimal conversion of all numbers.

First, add 101101 to 111010 as follows: write one number beneath the other and commence adding at the last column on the right side.

Col. 7	Col. 6	Col. 5	Col. 4	Col. 3	Col. 2	Col. 1
	1	0	1	1	0	1
	1	1	1	0	1	0
1	1	0	0	1	1	1

For columns 1, 2 and 3 we have $1 + 0$ or $0 + 1$ giving a digit 1 in each column. Column 4 is $1 + 1$ giving 0 in the answer, carry 1 to column 5. Column 5 becomes $0 + 1 + 1$ (carried) giving 0 in column 5 answer, carry 1 to column 6. Column 6 becomes $1 + 1 + 1$ (carried) giving 1 in column 6 answer, carry 1 to column 7 answer. (The above layout is for convenience of explanation and need not be followed once the basic principles are understood.) Now add the answer 1100111 to the remaining number 1011 using the routine outlined below.

```
1100111
  1011
-------
1110010
```

Column 1 $1 + 1 = 0$ carry 1 to column 2
Column 2 $1 + 1 + 1$ (carried) $= 1$ carry 1 to column 3
Column 3 $1 + 0 + 1$ (carried) $= 0$ carry 1 to column 4
Column 4 $0 + 1 + 1$ (carried) $= 0$ carry 1 to column 5
Column 5 $0 + 1$ (carried) $= 1$
Column 6 $1 + 0$ (understood) $- 1$
Column 7 $1 + 0$ (understood) $= 1$

The final answer is thus 1110010, which may be checked by conversion as follows

101101 in decimal means $32 + 8 + 4 + 1$ that is, 45
111010 in decimal means $32 + 16 + 8 + 2$ that is, 58
 1011 in decimal means $8 + 2 + 1$ that is, 11

The answer 1110010 in decimal means $64 + 32 + 16 + 2$, that is, 114: since $45 + 58 + 11 = 114$ the answer is correct.

Subtraction

A set of basic rules for binary subtraction may be derived in a similar form to those derived for binary addition. However, since these involve 'borrowing' 1 from another column whenever 1 is subtracted from 0 this subtraction process is often found confusing. A better method is to use the addition process 'in reverse' by determining what number must be added to the number that is to be subtracted in order to equal the number from which that subtraction is to take place. To clarify this consider, for example, subtracting 1111 from 11001; the process reduces to finding the number which must be added to 1111 to equal 11001.

Example 14.16

Subtract 1111 from 11001.

	Column 5	Column 4	Column 3	Column 2	Column 1
Row 1	1	1	0	0	1
Row 2		1	1	1	1
Row 3	0	1	0	1	0

In column 1 consider what number must be added to the 1 in the number being subtracted (row 2) to equal the 1 in the number from which the subtraction is to take place (row 1). The answer is 0 which is placed in column 1, row 3.

Next consider column 2. What number must be added to the 1 in row 2 to equal the 0 in row 1? Since $1 + 1 = 0$ carry 1, the answer is 1, which is put in column 2, row 3.

In column 3 row 2 we now have 1 plus 1 carried from column 2. What number must be added to this $1 + 1$ to give 0 in column 3, row 1? The answer is 0 since $1 + 1 = 0$ carry 1 (the 1 to be carried to column 4). Write 0 in column 3, row 3.

In column 4, row 2 we now have $1 + 1$. What number must be added to the $1 + 1$ in row 2 to give the 1 in row 1? Clearly this must be 1 since $1 + 1 + 1 = 1$ carry 1 (the 1 to be carried to column 5). Write 1 in row 3, column 4.

Finally, in column 5, row 2 we now have a 1 carried from column 4. What number must be added to 1 to give the 1 in row 1? The answer is 0. Write 0 in row 3, column 5.

The overall answer is thus 01010 or 1010 since a 0 is the first digit and may be omitted. Check by adding together rows 2 and 3.

	Row 2		Row 3		Carried		Row 1	Carry	
Column 1	1	+	0			=	1		
Column 2	1	+	1			=	0	1	to column 3
Column 3	1	+	0	+	1	=	0	1	to column 4
Column 4	1	+	1	+	1	=	1	1	to column 5
Column 5			0	+	1	=	1		(complete)

The student is advised to read through this example again and ensure each line is fully understood. More examples with reducing detailed explanation follow.

Example 14.17

Subtract 11011 from 111010.

$$\begin{array}{r} 111010 \\ 11011 \\ \hline 11111 \\ \hline \end{array}$$

Column 1 (right-hand side)
>To give 0 in row 1, 1 must be added to the 1 in row 2. Write 1 in row 3. This will carry 1 to column 2 when rows 2 and 3 are added.

Column 2 To give 1 in row 1, 1 must be added to the 1 in row 2 and the 1 from column 1. Write 1 in row 3. This will carry 1 to column 3 when rows 2 and 3 are added.

Column 3 To give 0 in row 1, 1 must be added to the 1 in row 2 carried from column 2. Write 1 in row 3. This will carry 1 to column 4 when rows 2 and 3 are added.

Column 4 To give 1 in row 1, 1 must be added to the 1 in row 2 and the 1 carried from column 3. Write 1 in row 3. This will carry 1 to column 5 when rows 2 and 3 are added.

Column 5 As column 4. Write 1 in row 3. This will carry 1 to column 6. Row 1 column 6 is 1. No further digits needed.

Subtracting in denary gives $58 - 27$ which is 31. Binary equivalent of 31 is 11111, thus the answer is correct.

Example 14.18

Subtract 110110 from 111000.

$$\begin{array}{r} 111000 \\ 110110 \\ \hline 000010 \\ \hline \end{array}$$

Example 14.19

Subtract 111011 from 1110000.

$$\begin{array}{r} 1110000 \\ 111011 \\ \hline 110101 \\ \hline \end{array}$$

Approximations

Two phrases are of particular importance when using decimal notation. These are 'places of decimals' and 'significant figures'. The number of places of decimals or significant figures in an answer of the kind found in the example 14.14 determine its accuracy. When such an answer is reduced to a few figures an approximation has been made.

The phrase 'places of decimals' is self explanatory and refers to the number of figures following the decimal point. For example, consider the number (written to four places of decimals) 3.1428. If we wish to write this to three places we have to decide what the third figure following the point should be to obtain a reasonable degree of accuracy. Since 3.1428 is closer in value to 3.143 than 3.142 then 3.143 would be the correct answer to three places. In general the rule is: if the figure following the figure in the last place required lies between 5 and 9, the last figure is increased by 1; if not it remains as it is.

Example 14.20

Express the following decimal numbers to the number of places of decimals shown.

(a) 53.678 to 2 places   (b) 1.78 to 1 place
(c) 1.597 to 2 places   (d) 1.594 to 2 places

(a) The figure in the third place is 8, change the preceding 7 to 8—answer is 53.68.

(b) The figure in the second place is 8, change the preceding 7 to 8—answer is 1.8.

(c) The figure in the third place is 7, change the preceding 9 to 10 which in turn changes the 5 to 6—answer is 1.60.

(d) The figure in the third place is 4. Leave the preceding 9 at 9—answer is 1.59.

The phrase 'significant figures' means the number of figures other than zero which in total make up the number, regardless of whether the figures lie before or after the point. The same rule of approximation applies as above. For example

19.35 to 3 significant figures 19.4
19.35 to 2 significant figures is 19

1.985 to 3 significant figures is 1.99
1.985 to 2 significant figures is 2.0

1273.5 to 3 significant figures is 1270
1273.5 to 2 significant figures is 1300

Standard Form—Powers of Ten

Writing a decimal number in standard form means having only one figure before the decimal point and adjusting the overall value to that required using a multiplier of ten raised to the appropriate power.

Example 14.21

Express in standard form

   (a) 198.3   (b) 17.53   (c) 5726.5   (d) 0.534

   (a) $198.3 = 1.983 \times 100$ that is 1.983×10^2
   (b) $17.53 = 1.753 \times 10$
   (c) $5726.5 = 5.7165 \times 1000$ that is 5.7265×10^3
   (d) $0.534 = 5.34/10$ that is 5.34×10^{-1}

Standard form is particularly useful when estimating an answer. For example, evaluate

$$\frac{19.7 \times 1.8 \times 5613}{4.9 \times 35.1}$$

This can be written

$$\frac{1.97 \times 10 \times 1.8 \times 5.613 \times 10^3}{4.9 \times 3.51 \times 10}$$

that is

$$\frac{1.97 \times 1.8 \times 5.613}{4.9 \times 3.51} \times 10^3$$

when tens are cancelled. This is approximately

$$\frac{2 \times 2 \times 6}{5 \times 4} \times 10^3$$

that is

$$\frac{24000}{20} \text{ or } 1200$$

(The actual answer is 1157.26.)

   Estimation of this sort is usually used when the order of magnitude of an answer is required. Often the actual figures (in this case 1157.26) are determined by some other means, for example a slide rule or logarithms. A calculator, of course, gives the position of the decimal point automatically.

   Powers of 10 are commonly used as a multiplier to reduce or increase a number to a reasonable size. When used, a prefix is attached to the unit of whatever is being measured to indicate the power of 10. A list of prefixes and their abbreviations is given in table 14.1.

Multiplier	Prefix	Abbreviation	Multiplier	Prefix	Abbreviation
10^{12}	tera	T	10^{-12}	pico	p
10^9	giga	G	10^{-9}	nano	n
10^6	mega	M	10^{-6}	micro	
10^3	kilo	k	10^{-3}	milli	m
10^2	hecto	h	10^{-2}	centi	c
10	deca	da	10^{-1}	deci	d

table 14.1

For cxample:

A current of 0.0135 amperes is expressed as 13.5 milliamperes (abbreviated mA) since $0.0135 = 13.5 \times 10^{-3}$.

A voltage of 5673 volts is expressed as 5.673 kilovolts (abbreviated kV) since $5673 = 5.673 \times 10^3$.

A resistance of 5 600 000 ohms is expressed as 5.6 megohms (abbreviated MΩ) since $5\ 600\ 000 = 5.6 \times 10^6$.

In cases like this it is usual to stop dividing when a sufficiently accurate answer is obtained.

The units ampere, volt and ohm and the quantities current, voltage and resistance are discussed elsewhere in the book.

Conversion of Proper Fractions to Decimal Fractions

Long or short division may be used to convert a proper fraction to a decimal fraction. In example 14.14 we saw 22/7 converted to 3.1428 etc. Similarly

$$\frac{3}{7} = 0.428 \text{ etc.}$$

by dividing 7 into 3.000.

$$\frac{2}{5} = 0.4$$

by dividing 5 into 2.0.

$$\frac{5}{8} = 0.625$$

by dividing 8 into 5.000.

Also

$$3^3/_7 = 3.428$$

either by writing 3³/₇ as 24/7 and dividing 24.00 by 7 or by putting the whole number 3 before the point of the decimal fraction 0.428 (which equals 3/7). Similarly

$$4²/₅ = 4.4$$

either by evaluating 22/5 or placing 4 before 0.4, and

$$3⁵/₈ = 3.625$$

either by evaluating 29/8 or by placing 3 before 0.625.

Conversion of Decimal Fractions to Proper Fractions

Example 14.22

Convert the following to proper fractions or mixed numbers as appropriate (a) 0.76   (b) 0.125   (c) 3.5   (d) 56.625   (e) 1.28

(a) $0.76 = \dfrac{76}{100}$

$= \dfrac{19}{25}$ (dividing through by 4)

(b) $0.125 = \dfrac{125}{1000}$

$= \dfrac{1}{8}$ (dividing through by 125)

(c) $3.5 = \dfrac{35}{10}$

$= \dfrac{7}{2}$ (dividing through by 5)

$= 3½$

(d) $56.625 = 56 + 0.625$

$= 56 + \dfrac{625}{1000}$

$= 56 + \dfrac{5}{8}$ (dividing fraction through by 125)

$= 56⁵/₈$

(e) $1.28 = \dfrac{128}{100}$

$= \dfrac{32}{25}$ (dividing through by 4)

$= 1^{7}/_{25}$

Percentages

The numerator of a fraction containing 100 as the denominator is called a percentage. Thus for example 43 per cent (written 43%) means 43/100. A percentage figure is usually used to indicate a part of a whole, as for example in the statement '43 per cent of the students passed the examination'. This means 43/100 of the number of students taking the examination were successful. Thus if 100 students took the examination

$\dfrac{43}{100}$ of 100 that is $\dfrac{43}{100} \times 100$

or

43 students passed.

Similarly, if 200 students took the examination and 43 per cent passed, then

$\dfrac{43}{100} \times 200$

that is 86 students were successful.

Any fraction, proper, improper or decimal may be converted to a percentage.

Example 14.23

Convert the following to percentages: (a) 2/5   (b) 3/4   (c) 0.47 (d) 0.073   (e) 3/7

(a) $\dfrac{2}{5} = \dfrac{40}{100}$ (multiplying through by 20)

Thus $\dfrac{2}{5}$ of any number means 40 per cent of that number.

(b) $\dfrac{3}{4} = \dfrac{75}{100}$ (multiplying through by 25)

$\quad\quad = 75\%$

(c) $0.47 = \dfrac{47}{100}$

$\quad\quad = 47\%$

(d) $0.073 = \dfrac{73}{1000}$

$\quad\quad = \dfrac{7.3}{100}$

$\quad\quad = 7.3\%$

(e) Since 100 is not exactly divisible by the denominator 7 convert to a decimal fraction

$\dfrac{3}{7} = 0.4285$ (to 4 places)

$\quad = \dfrac{42.85}{100}$

$\quad = 42.85\%$

Formulae and Equations

Algebra is a means of referring to physical quantities using symbols instead of writing the name or the quantity in full. Thus instead of writing the basic relationship between voltage, current and resistance, which is

$$\frac{\text{voltage}}{\text{current}} = \text{resistance}$$

we might write

$$\frac{V}{I} = R$$

where

V represents voltage (measured in volts)
I represents current (measured in amperes)
R represents resistance (measured in ohms)

This relationship is called an equation. The expression V/I could be referred to as the formula for resistance. In this section we shall be further examining formulae (note the plural) and equations, in particular looking at the process of transposition when it is desired to change the subject of a formula and also at the solution of equations, that is finding the value of a variable (the unknown) under circumstances where all other determining factors are known.

Transposition

In the formula $R = V/I$ given above, the resistance R is the subject of the formula since its value is given in terms of the values of voltage V and current I at any particular time.

For example, if $V = 10$ V and $I = 5$ A then R is 10/5, that is, 2 Ω. Arranged in this way it is easier and more convenient to find R for all the different values that V and I may have. However, in many circumstances the voltage and resistance may be known, in which case it may be required to find the value of the current. The formula now required is of the form in which current I lies on the left-hand side. By 'appropriately arranged' we mean that they must be arranged in a way which does not alter the scientific truth of the statement that resistance is equal to voltage divided by current. This rearrangement is called *transposition* of the formula—in this case making I the subject. Transposition must be carefully handled since any error will result in making a statement which is not true. In engineering the use of an untrue mathematical statement could at best lead to a loss of expensive equipment or components, and at worst the loss of human life. It is therefore essential never to make changes in an equation without ensuring that the original truth is unaltered.

A formula or equation giving the relationship between one variable and other variables may be likened to a balance in which masses are compared when the masses are equal. Anything which is done to one side of the balance must be done to the other or the state of balance is changed. With an equation or formula this would mean that the original equality (implied by the sign =) is no longer valid if one side of the equation is changed differently to the other.

To return to the formula $R = V/I$. To make I the subject multiply both sides by I giving

$$IR = \frac{VI}{I}$$

thus $$IR = V$$

(since I divided by I is unity).

Now divide both sides by R

$$\frac{IR}{R} = \frac{V}{R}$$

and so

$$I = \frac{V}{R}$$

To make V the subject

$$R = \frac{V}{I}$$

multiply both sides by I as before

$$IR = \frac{VI}{I}$$

that is

$$V = IR$$

(Note that the sides may be interchanged from left to right provided that they are not altered in any way in so doing.)

Study the following examples carefully.

Example 14.24

The relationship between electrical resistance R, resistivity ρ, conductor length l and conductor cross-sectional area A is given by

$$R = \rho \, \frac{l}{A}$$

Make the resistivity ρ the subject of the formula.

Multiplying both sides by A

$$AR = \rho \, \frac{lA}{A}$$

$$= \rho \, l$$

Divide both sides by l

$$\frac{AR}{l} = \rho \, \frac{l}{l}$$

$$= \rho$$

thus

$$\rho = \frac{AR}{l}$$

that is

$$\text{resistivity} = \frac{\text{area} \times \text{resistance}}{\text{length}}$$

Example 14.25

The velocity of a moving body v after a time t spent accelerating with a constant acceleration a is given by

$$v = u + at$$

where u is the velocity of the body at the beginning of time t. Make t the subject of this formula.

Subtract u from both sides

$$v - u = u - u + at$$

$$= at$$

Divide both sides by a

$$\frac{v - u}{a} = \frac{at}{a}$$

$$= t$$

Thus

$$t = \frac{v - u}{a}$$

that is

$$\text{time} = \frac{\text{final velocity} - \text{initial velocity}}{\text{acceleration}}$$

From the above examples some basic rules of transposition can be seen. These are: first isolate the term containing the required subject on one side of the equation (in example 14.25 by subtracting u from both sides); second isolate the subject. If the subject is contained in the numerator of a fraction then divide through by any multiplier in the same numerator (dividing by l in example 14.24 and by a in example 14.25) and multiply through by the denominator of the fraction containing the subject (A in example 14.24, 'denominator' of unity in example 14.25 and thus multiplication is not necessary).

Example 14.26

Make *a* the subject of

$$y = \frac{ax + b}{c}$$

Multiply through by *c*

$$cy = ax + b$$

Subtract *b* from both sides

$$cy - b = ax$$

Divide through by *x*

$$\frac{cy - b}{x} = a$$

Thus

$$a = \frac{cy - b}{x}$$

Example 14.27

Make *x* the subject of

$$y = x^{1/2} + 1$$

($x^{1/2}$ means the square root of *x*, that is the number which when multiplied by itself equals *x*).

Subtract 1 from both sides

$$y - 1 = x^{1/2}$$

Square both sides of the equation

$$(y - 1)^2 = (x^{1/2})^2$$

Thus

$$(y - 1)^2 = x^{1/2 \times 2} \text{ (multiple powers)}$$

$$= x$$

and

$$x = (y - 1)^2$$

or

$$x = y^2 - 2y + 1$$

Note a new aid to transposition is used here, the squaring of both sides of an equation to remove the square root. This is quite in order since if two quantities are equal then clearly their squares must be equal.

Example 14.28

Make R_2 the subject of

$$R_T = \frac{R_1 R_2}{R_1 + R_2}$$

Multiply both sides by $(R_1 + R_2)$

$$R_T(R_1 + R_2) = R_1 R_2$$

The required subject is now on both sides of the equation. Terms containing R must be gathered together on one side. Expand the bracket to obtain the term containing R_2 on the left-hand side

$$R_T R_1 + R_T R_2 = R_1 R_2$$

Subtract $R_T R_2$ from both sides

$$R_T R_1 = R_1 R_2 - R_T R_2$$

Use brackets to extract R_2

$$R_T R_1 = (R_1 - R_T) R_2$$

Divide through by $(R_1 - R_T)$

$$\frac{R_T R_1}{R_1 - R_T} = R_2$$

Evaluation of Formulae

Having obtained the formula with the required subject it may then be necessary to find the value of the variable which is the subject for given values of the other variables. Sometimes in such cases the subject variable is referred to as the dependent variable (since it depends on the values of the other variables) and the other variables as the independent variables.

Thus, in the formula $R_T = R_1 R_2/(R_1 + R_2)$ for example (in which R_T represents total resistance of two resistors having resistance R_1 and R_2 connected in parallel), R_T is the dependent variable and R_1, R_2 independent variables.

Evaluation of a formula is fairly straightforward and all that is required is to write in the given values of the variables and calculate the value of the subject using logarithms, slide rule or calculator. In most practical examples multiples or sub-multiples of units using powers of 10 are

involved, for examples milliamps, microvolts etc. and care must be taken
to insert the correct multiplier. Care must also be taken to ensure that in
engineering formulae the correct consistent units are used throughout.

Example 14.29

The value of the resultant resistance R_T when two resistors having
resistance R_1, R_2 are connected in parallel is

$$R_T = \frac{R_1 R_2}{R_1 + R_2}$$

where R_T, R_1 and R_2 are all measured in ohms.

Calculate R_T to three significant figures when $R_1 = 1\ k\Omega$ and $R_2 = 1200\ \Omega$

$$R_T = \frac{1000 \times 1200}{1000 + 1200} \quad (\text{writing } 1\ k\Omega \text{ as } 1000\ \Omega)$$

$$= 545\ \Omega$$

Example 14.30

The periodic time of a pendulum of length l metres is given by

$$T = 2\pi \sqrt{\frac{l}{g}} \text{ seconds}$$

where g is the acceleration due to gravity (9.81 m/s²) and π is a constant
equal to 3.14. Find T when $l = 25$ cm.

As stated, when using the formula l must be in metres. Now $l = 25$ cm,
that is 25×10^{-2} m or 0.25 m. Thus

$$T = 2 \times 3.14 \sqrt{\frac{0.25}{9.81}}$$

$$= 1.002\ s$$

Example 14.31

The power P in an electrical circuit in terms of the circuit resistance R and
the current I is given by the equation $P = I^2 R$ where P is measured in
watts, I in amps and R in ohms. Calculate the current in milliamps which
flows in a circuit of resistance 4.7 $k\Omega$ when the power is 500 mW.

First we must make I the subject of the formula $P = I^2R$. Divide by R on both sides

$$\frac{P}{R} = I^2$$

Take the square root of each side

$$I = \sqrt{\frac{P}{R}}$$

Now

$$P = 500 \text{ mW which is } 0.5 \text{ W}$$

and

$$R = 4.7 \text{ k}\Omega \text{ which is } 4700 \ \Omega$$

Thus

$$I = \sqrt{\frac{0.5}{4700}}$$

$$= \sqrt{\frac{5000}{4700}} \times 10^{-4}$$

Note the use of the multiplier 10^{-4} to make the placing of the decimal point easier.

$$= \sqrt{(1.063 \times 10^{-4})}$$

$$= 1.03 \times 10^{-2}$$

(Find the square root of 1.063 and halve the index of the multiplier.)

$$= 10.3 \times 10^{-3}$$

(Move the decimal point to obtain the right multiplier for mA.)

$$I = 10.3 \text{ mA}$$

Graphs

In engineering there are many situations in which the value of one quantity determines the value of another, the relationship between the two quantities being described by an equation. In the cases where there is only one equation and two variable quantities there are many sets of values which satisfy the equation and it is useful to show these values in pictorial form using a graph. The shape of the graph is determined by the equation and more often than not the process works the other way round, in that from the shape of a given graph obtained by experimental observation the equation relating the variable quantities may be deduced. This can then be used to predict values of the variables in a situation which has not actually been observed. This fact is especially useful in cases where such observations might be dangerous either to equipment or personnel or both.

Consider the equation $y = 2x$. In this equation y is the subject and is called the dependent variable, x is called the independent variable. In any such equation where y is the dependent variable y is said to be a function of x since the value of y is determined by the value of x. Similarly, in $a = 3b + 4$, a is a function of b or in $p = 3q - 2$, p is a function of q. Having looked at the terms used, now let us examine the techniques used in drawing graphs.

Some values of x and y which satisfy the equation $y = 2$ are as follows:

y	2	4	6	8	10	0	−2	−4	−6	−8	−10
x	1	2	3	4	5	0	−1	−2	−3	−4	−5

Such an arrangement is called a table of values. In practical examples the table should show, next to the variable, what units if any the variable is being measured in. For this equation the graph must be able to show that when x is 1, y is 2, when x is 2, y is 4, when x is −1, y is −2 and so on. We must therefore have some means of showing both positive and negative values of both variables.

The usual arrangement is shown in figure 14.1. Two lines called *axes* are drawn, one vertical and one horizontal, their point of intersection

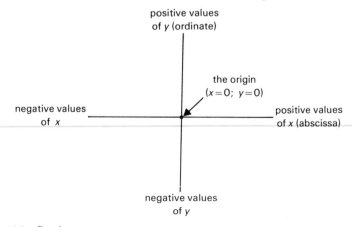

figure 14.1   Graph axes

being called the *origin*. Values of one variable are marked along one axis, values of the other being marked on the other axis. The variable marked along the horizontal axis is called the *abscissa* and that marked on the vertical axis is called the *ordinate*. Positive values of the abscissa are marked to the right of the origin, negative values to the left. Positive values of the ordinate are marked above the origin, negative values below. When x and y are used as symbols for variable quantities y is usually (but not always) taken as ordinate and x as abscissa.

As shown the graph area is divided into four quarters. These are called *quadrants*. In the upper right-hand quadrant values of both variables are positive, in the lower left-hand quadrant they are negative. In the upper left-hand quadrant the ordinate variable has positive values and the abscissa variable negative values and in the lower right-hand quadrant the ordinate has negative values and the abscissa positive. Not all quadrants need to be drawn depending on the values that are to be plotted; for example if only positive values of both variables are to be plotted then only the upper right-hand quadrant need be drawn. When all four quadrants are used the point where the axes cross one another—the origin—is the point where both variables have the value zero. If only one quadrant is used this need not necessarily be so, depending on what values are to be plotted. (There would be little point for example, in starting the scale at $x = 0$ if the given values lay between 200 and 250.)

The marking of the single point on the graph corresponding to each set of values of the variables is called plotting the values. Now examine figure 14.2 which plots the graph of $y = 2x$.

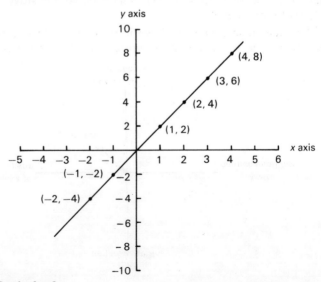

figure 14.2   Graph of $y=2x$

When $x = 0$, $y = 0$. One point on the graph is therefore the origin. When $x = 1$, $y = 2$. Where the vertical line running through $x = 1$ meets the horizontal line running through $y = 2$ is the point on the graph at which $x = 1$ and $y = 2$.

Note that graphs are usually plotted on special paper on which is printed a series of vertical and horizontal lines to help make placing the point easier. Not all such lines will be shown in the diagrams in this section in the interests of clarification of the diagrams.

Similarly, where the vertical line through $x = 2$ meets the horizontal line through $y = 4$ is the point on the graph where $x = 2$, $y = 4$.

Examine the graph for the remaining points and note how such a point may be indicated by writing the values of the variables in brackets with the abscissa value first. (It is not necessary to do this on an engineering graph. Such a practice would overfill the graph with figures and reduce the clarity. It is used here for demonstration.)

As can be seen, the graph of $y = 2x$ is a straight line and as such needs only two points to locate it. In fact all graphs of linear equations are straight lines and this is the reason for the use of the word linear.

The reader is advised to plot the graphs of the following equations, which will be discussed in some detail. The table of values is given in table 14.2 and figure 14.3 shows the shape of the graphs.

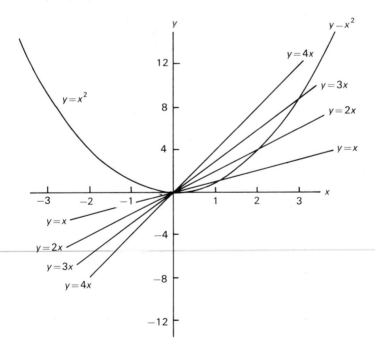

figure 14.3   Graphs of $y=x$, $y=2x$, $y=3x$, $y=4x$ and $y=x^2$

x	-3	-2	-1	0	1	2	3
$y = x$	-3	-2	-1	0	1	2	3
$y = 2x$	-6	-4	-2	0	2	4	6
$y = 3x$	-9	-6	-3	0	3	6	9
$y = 4x$	-12	-8	-4	0	4	8	12
$y = x^2$	9	4	1	0	1	4	9

table 14.2

From an examination of figure 14.3 the following points can be made:

(1) The graph of $y = x^2$ is not a straight line but is a curve symmetrical about the y axis and passing through the origin.

(2) All the remaining graphs are straight lines passing through the origin.

(3) The slope of $y = 4x$ is greater than that of $y = 3x$, which is greater than that of $y = x$.

The Straight Line Graph

All equations containing an independent variable raised only to a power 1, that is if the independent variable is x, and there are no terms containing x^2, x^3 etc., produce straight lines when plotted as graphs. The important characteristics of such graphs are first the *slope* or *gradient* of the graph, and second whether or not the graph passes through the origin and if it does not, where it crosses the ordinate axis.

The Gradient

The gradient or slope of any line is the ratio between the vertical rise or fall of the line from the horizontal over a measured horizontal distance to the value of this measured distance. Thus if for every 5 cm measured along the horizontal a line rises a vertically measured distance from the horizontal of 1 cm as shown in figure 14.4a then the gradient is said to be 1/5. If the line were to fall in a similar manner the gradient would then be taken as negative and equal to $-1/5$ (figure 14.4b).

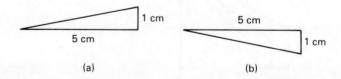

(a)   (b)

figure 14.4   (a) Positive gradient +1/5 (a similar system is used when measuring road gradients but here the ratio is between the rise or fall of the road per distance measured along the road. Because of the relatively small gradients this latter distance is approximately equal to the distance travelled horizontally

Consider the straight-line graphs shown in figure 14.3. Here the horizontal distance is a change in the value of x and the vertical rise or fall is the corresponding change in the value of y. Thus, for the graph of $y = x$ as x changes from $x = 0$ to $x = 3$, y changes from $y = 0$ to $y = 3$ and the ratio of vertical rise : horizontal distance equals 3 : 3 or 1—the gradient is unity.

Similarly for $y = 2x$ as x changes from 0 to 3, y changes from 0 to 6 and the gradient is 6/3 or 2. For $y = 3x$, the change in y of 9 (from $y = 0$ to $y = 9$) corresponds to a change in x of 3 (from $x = 0$ to $x = 3$) and the gradient is 9/3 or 3. Finally for $y = 4x$ the gradient is 12/3 or 4.

The gradient in each case is the coefficient of x and in general for equations of the form $y = mx$ (or $a = mb$ or $c = md$) where m is a constant and the other symbols represent variable quantities then m is the gradient of the graph.

Example 14.32

Find the gradients of the graphs of the following equations: (a) $y = 5x$ (b) $4y = 3x$ (c) $2y = -x$ (d) $7a = -6b$.

(a) The gradient is 5.
(b) Rearrange $4y = 3x$ to give $y = \frac{3}{4}x$. The gradient is 3/4.
(c) Rearrange $2y = -x$ to give $y = -x/2$ (that is the coefficient of x is $-1/2$). The gradient is $-1/2$.
(d) Rearrange $7a = -6b$ to give $a = -6b/7$. The gradient is $-6/7$.
In examples 14.32c and 14.32d the graphs slope downwards.

The Intercept

The value of the variable at the point where the graph cuts one or other of the axes is called the intercept of the graph on the axis concerned. Thus if $x = 6$ at the point where $y = x - 6$ cuts the x axis then the intercept on the x axis is 6.

The value of the *intercept* on either axis depends on the equation and on the situation of the axes relative to one another. In graphs using all quadrants the y (ordinate) axis crosses the x (abscissa) axis where $x = 0$ and similarly the x axis cross the y axis where $y = 0$. The point of crossing, called the origin, is thus $x = 0$, $y = 0$. To find the intercept under these conditions all that is necessary is to put $x = 0$ in the equation to find the intercept on the y axis and $y = 0$ in the equation to find the intercept on the x axis.

If either axis crosses the other at any other point, that is if the scale variable does not start at zero, then the intercept will be different and is not the value determined by putting $x = 0$ or $y = 0$ into the equation.

Consider the graphs of the following functions

$$y = x + 1, \quad y = x + 3, \quad y = x - 1$$

The table of values is:

x	-2	-1	0	1	2
$y = x + 1$	-1	0	1	2	3
$y = x + 3$	1	2	3	4	5
$y = x - 1$	-3	-2	-1	0	1

table 14.3

(Only two values of x need be taken, the others are included as an additional emphasis that the graphs are straight lines.)

The shape and position of these graphs relative to one another is shown in figure 14.5. As can be seen the graph of $y = x + 1$ cuts the y axis at $y = 1$, the graph of $y = x + 3$ cuts the y axis at $y = 3$ and the graph of $y = x - 1$ cuts the y axis at -1. Thus the intercepts on the y axis when $x = 0$ of these

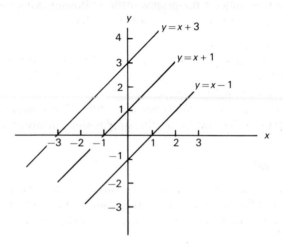

figure 14.5

three graphs are 1, 3 and -1, respectively. Look at the values for $x = 0$ in the table above; this column gives the value of the intercepts of the three graphs. Thus by putting $x = 0$ into an equation relating y and x the intercept on the y axis may be found.

Similarly the intercept on the x axis, that is where the graph cuts the x axis, may be found by putting $y = 0$. These intercepts are -1, -3 and 1 for the three graphs as shown in the figure.

Appendix 1: Health and safety

Millions of man-hours and many lives are lost every year through accidents at work. Most, if not all, of these accidents need never happen if everyone at work kept safety uppermost in their minds at all times. A moment's thoughtlessness can result in permanent injury or even death and everyone has a responsibility to themselves and, equally importantly, to others to use safe working practices and to behave sensibly and responsibly at all times.

Parliament has played its part in trying to protect the health and safety of persons at work and the Factories Act and the Offices, Shops and Railway Premises Act lay down rules and regulations and procedures to be adopted in places of work. More recently the extent to which the law governs working conditions and the behaviour of employers and employees has been considerably extended by the Health and Safety at Work Act 1974.

The Health and Safety at Work Act 1974

The Factories Act and the Offices, Shops and Railway Premises Act mentioned above did not cover all places of work and, many thought, did not go far enough in promoting the health and safety of people at work and people who may be affected by work going on near by. The 1974 Act extended the power of the law in a number of ways and for the first time about five million people who work in education, research, medicine, the leisure industries and certain parts of the transport industry received protection. Further, the new Act extended this protection to others who may be affected by work taking place near by and, secondly, placed a definite responsibility for health and safety on the employee as well as the employer. Careless or irresponsible behaviour of an employee which leads or may lead to accidents is now an offence and prosecutions have taken place under the 1974 Act.

The aim of the 1974 Act is to

'secure the health, safety and welfare of persons at work; protect persons other than persons at work against risks to health or safety arising out of or in connection with activities of persons at work; control the keeping and use of explosive or highly flammable or otherwise dangerous substances and to prevent the unlawful acquisition of such substances and control the emission into the atmosphere of noxious or offensive substances.'

(Health and Safety Commission Leaflet HSC3)

The duties of employers are laid out in Section 2 of the Act. Briefly, they are to ensure as far as is possible the health, safety and welfare of employees. Employers must provide and maintain plant and systems of work that are safe and without risks to health, must make arrangements that ensure safety and absence of risk in the use, handling, storage and transport of materials, must provide instruction, training and supervision as is necessary to ensure as far as possible the health and safety of employees, must provide safe access to and from places of work and properly maintain the working environment. Section 3 of the Act places similar responsibilities on the employer towards those who are not his employees but may be affected by the work being done.

Section 4, 5 and 6 of the Act define the responsibilities of people who own and let premises for work, people whose premises may emit 'noxious or offensive substances' and those who design, manufacture, import or supply any article for use at work. Again, in general, the duties are to ensure as far as possible the health, safety and welfare of those affected either at work or by work being carried on near by.

A most important part of the Act as far as the employee is concerned is contained in Section 7 and Section 8. These sections which are new in health and safety legislation define the duties of the employee. Section 7 states

'It shall be the duty of the employee while at work: to take reasonable care for the health and safety of himself and of other persons who may be affected by his acts or omissions at work, and, as regards any duty or requirement imposed on his employer or any other person by or under any of the relevant statutory provisions, to co-operate with him so far as is necessary to enable that duty or requirement to be performed or complied with.'

Section 8 of the Act states that

'no person shall intentionally or recklessly interfere with or misuse anything provided in the interests of health, safety or welfare in pursuance of any of the relevant statutory provisions'.

In summary then, the 1974 Act which came into force in full by April 1975 places responsibility for health and safety for persons at work or persons who may be affected by work being done near by on both employer and employee. However, while Acts of Parliament may encourage and reinforce proper standards of behaviour and practice they cannot guarantee that they are followed at all times.

Each individual should be fully aware of safe and unsafe practice in his own trade and should always be watching for potential hazards which may cause an accident to himself or others. The rest of this appendix describes a number of safe and unsafe practices in general and in electrical and electronic trades in particular.

General Safety

A phrase often used in safety and accident prevention guides is 'good housekeeping'. Simply it means *tidiness*, which is essential in any kind of practical work. The following five simple rules for good housekeeping should be remembered and put into practice always

1. Working surfaces should be kept clean and uncluttered.
2. Aisles and gangways should be kept clear at all times.
3. Floors should be clean, not slippery and in good repair.
4. Waste bins for materials no longer of any use should be available and should be marked clearly if only one kind of waste is to be put into them.
5. Jobs partially finished should be put to one side in a safe place so as not to interfere with the job in hand.

Accident prevention is also assisted by ensuring that you work in a well lit area (a minimum of 300 lux generally, with added illumination locally for detailed work) and that the working surface is at the correct height for comfort and ease of use. Remember it is your responsibility as well as that of your employer to seek and maintain, as far as you are able, a safe working environment.

Many jobs require protective clothing, which includes not only dress but additional items such as eye shields or glasses. When in doubt it is always wisest to protect yourself as far as possible from general and specific hazards of the work you are doing.

Again, if protective clothing or equipment is provided and not used and an accident occurs there will be reduced likelihood of compensation and the person concerned may even be liable to prosecution under the terms of the 1974 Act. Risk taking especially when there is the possibility, for example, of eye injury is never worth it and although certain of the

Protection of Eyes Regulations do not legally apply in all establishments, eye protection should be worn for all the processes indicated there. They include the more obvious ones such as milling, grinding and the use of abrasive wheels in general as well as processes involving chemical splashes such as electrical battery maintenance. There has in the past been insufficient emphasis on safety in certain areas and there is a tendency to believe that the use of protective clothing or equipment is 'soft' or 'silly' but the hard fact remains that limbs and eyes once permanently damaged are irreplaceable even with modern technology.

The biggest single reason for loss of man-hours in industry is back trouble. Cynics often say the reason for this is that no one can prove that a complainant does not have backache or pain if he says he has. This apart, however, genuine back problems abound and the basic reason in the first instance is incorrect lifting and carrying. The following general rules are useful and should be practised.

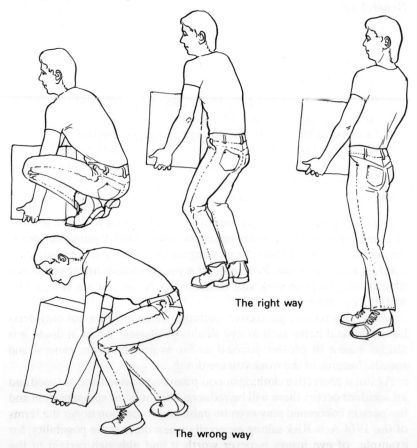

The right way

The wrong way

figure A1 Lifting correctly

1. Never attempt to lift or carry on your own a load which is excessively heavy, bulky or of a shape that one person cannot safely and conveniently handle. Get assistance.

2. When lifting use the leg muscles to take the strain rather than the back. This may be done by keeping the back straight, bending the knees and squatting downwards before attempting to lift. Never bend over to lift with the legs straight or partially straight. See figure A1.

3. Never carry a large object even if it is light if it obscures vision. You must at all times be able to see where you are going.

When stacking boxes or other items do so carefully so that there is no danger of them falling and remember 'good housekeeping'.

Electrical Safety

Electricity can kill. This simple and obvious statement should be borne in mind at all times. The passage of electric current through the body, particularly the nervous system, produces electric shock which may stop the heart and lungs and death by suffocation may occur if resuscitation is not carried out immediately. Serious injury may be caused by burns due to the destruction of skin and tissues by the passage of current through them.

The effect of electric shock depends upon a number of factors. These are voltage, current, the path through the body taken by the current and the surface resistance of the skin. Voltages as low as 15 V may produce slight shock and a voltage of 70 V has been known to cause death. Most fatalities occur, however, from the standard domestic and industrial voltage of between 230 V and 250 V a.c., the current range being between 25 mA and 30 mA. The current path is determined by which parts of the body come into contact with the conductors carrying the voltage, that is the *live* conductors, a path between hands, that is across the chest and via the heart and lungs, being the most dangerous.

It should be noted that with the standard methods of electricity distribution within the UK, which is discussed later, the live side of an a.c. supply is at a higher voltage than ground and it is necessary to touch only the live wire or conductors connected to it to receive a shock since electric current will flow through the body to ground.

The wearing of insulated sole shoes is recommended. Skin resistance depends particularly on dampness, perspiration or other liquid reducing the resistance and increasing the likelihood of death or serious injury.

Secondary effects from electric shocks, which do not in themselves kill, may occur from involuntary muscle contraction causing falls from platforms or ladders or injury to hands or arms as they are withdrawn from the electricity source. A common accident in maintenance and repair of television receivers is injury of this sort to the hand or arm as it catches the chassis, case or sharp metal components within the set as the arm is withdrawn hurriedly. Particular care must be taken in electrical and electronic maintenance and repair since, of necessity, fault finding work is carried out on live equipment. Certain television receivers have a chassis which is not at zero voltage and unless isolation transformers are available this is a potential hazard since it is a relatively large metal area which is live.

Mains electricity in the UK is usually alternating current generated at voltages of the order of 11 kV and transmitted at high voltages via the familiar landlines of the National Grid of up to 400 kV. At the receiving end the voltage is reduced via transformer sub-stations typically to 6600 V and then to 415 V or 240 V for industrial and domestic distribution.

The 415 V supply is *three phase*, which means there are three wires, often indicated in industrial installations by the colours red, yellow and blue and between any pair of wires the voltage is 415 V. A fourth wire called the *neutral* wire may also be present in a three phase, four wire system, the voltage between any single phase and neutral being 240 V. In domestic distribution one phase and the neutral are taken as the supply lines to premises, a third wire called the *earth* wire being connected to ground via water pipes or earthing rods. The method of connection of the neutral line normally ensures that it is at zero volts with respect to ground but it is sometimes possible for it to 'float', the voltage difference rising by as much as 100 V. The p.d. between each phase wire of a three phase

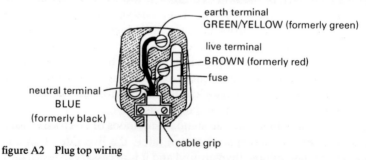

earth terminal
GREEN/YELLOW (formerly green)

live terminal
BROWN (formerly red)

fuse

neutral terminal
BLUE
(formerly black)

cable grip

figure A2   Plug top wiring

system (or the phase wire of a single phase system) and ground is the single phase voltage of 230 V–250 V. The phase wire is called the live wire and is denoted by L in plug tops and sockets, the neutral wire is denoted by N and the earth or ground wire by E. It is essential that plug tops and sockets are correctly wired as indicated in figure A2. The colour code in common use is:

LIVE—BROWN
NEUTRAL—BLUE
EARTH—GREEN AND YELLOW STRIPE

Earlier equipment may use red for live, black for neutral and green for earth.

All electrical supplies within buildings are controlled by appropriate switchgear capable of totally isolating circuits and are protected by fuses or circuit breakers, which, in the event of overload (that is, current being drawn beyond the safe value for the cables and loads), break the circuit from connection to the incoming supply. Before working on any equipment the location of the appropriate switches, fuses and/or circuit breakers must be known and if the switches are not within view of the equipment suitable notices must be attached to them to indicate work is being done on the circuit or equipment controlled by the switches. In some cases lockable switchgear is used and it is possible to secure the switches in the open or 'off' position. Switches should never be locked in the closed or 'on' position. When replacing fuses protecting circuits or equipment only the correct value must be used. A fuse of higher value allows the circuit to operate normally but the protection offered to the circuit or equipment is substantially reduced or may even be zero. Overloading may then cause fires and/or permanent damage.

A number of accidents involving electric shock occur because it is assumed that a circuit is isolated or 'dead' when it is not. Care should be taken before starting any work on any circuit or equipment to first

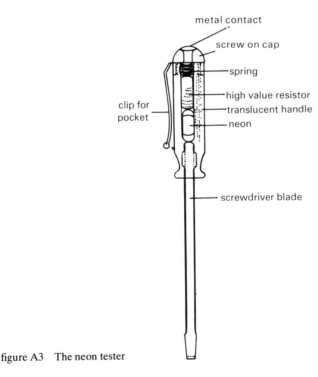

figure A3   The neon tester

determine which state it is in. In the case of a.c. mains supplies the phase or 'live' connection may be determined by using either a neon tester or test lamp. The neon tester, which is often in screwdriver form, consists of a blade, neon indicator and high value resistor in series. The thumb or finger is placed at one end, the other end being placed on a circuit or equipment part being tested. If the circuit is live a very small current sufficient to light the lamp but not sufficient to cause shock flows from the circuit through the blade, neon lamp and resistor, through the person testing to ground. The tester is perfectly safe up to the voltage indicated on it (several thousand volts). It should not be used on circuits or equipment carrying higher voltages than that given as the maximum safe value (see figure A3). Using a test lamp involves connecting one end to ground or neutral and connecting the other end to the point under test. If the lamp lights the circuit is obviously live. The test lamp method may be used on any supply, a.c. or d.c. provided that in the case of the latter the other connection of the lamp is made to the return cable. In all cases the lamp must be of a rating which will ensure that it will give an indication of the presence of a voltage. A neon tester will not normally work on a d.c. supply because there is no return path for the neon current.

As was stated earlier some types of work must be done on equipment when it is live and in this case extreme care must be taken at all times in attaching instruments or probes to the circuit.

The electrical connections to instruments and portable hand tools should be maintained in good condition at all times. Burnt or frayed cables on soldering irons or other equipment must be replaced as soon as the wear or damage is observed. Portable tools other than those known to be all-insulated or double-insulated should have a three core cable and the earth wire securely fixed to both tool and plug top.

All-insulated or double-insulated tools should have no touchable metal parts (other than perhaps the name plate, which, if it is metal, would be itself isolated from the supply) and must comply with British Standards (2769 or 3456) and bear the kite mark or the certified mark of the British Electrical Approvals Board. This type of equipment should be stored in a dry place and its insulation periodically checked. The supply cable to all- or double-insulated tools and equipment is two core (live and neutral) and there is no earth wire.

Regular inspection of all tools, test equipment, plugs, sockets and cables is essential. What was safe yesterday is not necessarily safe today.

Soldering has been described elsewhere. Safety rules to remember are

1. When not in use place the iron in a proper holder in the correct manner—not lying on the bench.

2. Never 'flick' or shake the iron to remove excess solder—wipe it; even small specks of molten solder can cause unpleasant injury to the unfortunate recipient.

Fire Safety

The importance of safety procedures to avoid fire cannot be overstressed; it is also essential to be familiar with the causes of fire and the methods to combat it should it arise. The damage and/or personal injury caused by a large number of fires every year in industry and the home could be avoided if prompt and correct action were taken immediately on discovery.

Three conditions are necessary for fire to start and keep going. These are fuel, oxygen and heat. Anything that will burn is regarded as a fuel in this sense and it should be remembered that apart from obvious materials such as wood, paper, fibres etc., even metals burn if they are finely divided. Oxygen, usually taken from the surrounding air, is an essential ingredient and certain materials such as celluloid, which gives off more oxygen as it burns, are particularly dangerous. Fire is a chemical reaction in the form of vapour burning; the vapour is produced by whatever constitutes the fuel being heated. Heat is thus the third ingredient. Flammable materials are those which will burn (note the word 'inflammable' is no longer used) and for each flammable material there is a minimum temperature at which the material will catch fire if a flame is near. Certain materials are dangerously flammable in that they catch alight of their own accord if the *self-ignition temperature* or *flash point* is reached.

To combat a fire once started it is necessary to remove one or more of the three ingredients. Usually, since the fuel cannot easily be removed, the technique is to starve the fire of oxygen and, if possible, to cool the burning material at the same time. There are seven main types of fire extinguisher. These are water, foam, dry powder, carbon dioxide, halons, blankets and dry sand. The basic principle of all of them is to remove one of the three conditions of fire, oxygen, although certain of the extinguishers, particularly water, help cool the burning material and reduce the effects of the second condition, heat.

Water may be used on all fires *except* electrical fires and flammable liquids. Clearly since water is a conductor (in its usual impure state) it is dangerous to spread it on fire in electrical equipment which may still be live. On burning flammable liquids water agitates the burning material and helps spread the fire further. Water, of course, causes its own kind of damage but this is of secondary importance when the prime consideration is to extinguish the fire.

The most common foam extinguisher contains two separate water solutions of aluminium sulphate and sodium bicarbonate. When the extinguisher is set off, usually by striking a plunger, the two solutions mix and a foam of aluminium sulphate, sodium sulphate and carbon dioxide is produced which is used to smother the fire and deprive it of the essential oxygen. The chemicals are not dangerous but should nevertheless be washed off skin and should not be allowed into contact with eyes or mouth. Foam extinguishers may be used on flammable liquids but not on electrical fires since, again, they contain water.

Dry powder extinguishers may be used on all fires, including electrical, although they may produce secondary damage to electrical and electronic equipment. The powder is usually very finely divided sodium bicarbonate.

Carbon dioxide extinguishers are particularly useful for electrical and flammable liquid fires. The suffocating property is excellent but since it has a very small cooling effect an extinguished fire may re-ignite. This type of extinguisher is the most common extinguisher provided in electrical and electronic workshops.

Fire extinguishers containing halons (vaporising liquids) are no longer in common use because their chemicals are poisonous.

Blankets and dry sand in buckets are often provided as a cheap and convenient method of extinguishing a fire.

A fire blanket may be used on burning liquids and most other fires small enough to be covered by a single blanket. They should be used with caution, care being taken to protect the hands and face from the heat.

Some blankets are made of asbestos, which is now regarded as hazardous in itself, or glass fibre which may also prove a danger to health. Aluminised glass fibre appears to be the safest choice. Dry sand is useful for extinguishing most fires but should not be used on liquid fires. Sand buckets should be checked regularly to ensure they contain clean, dry sand and they should *never* be used as litter bins.

Fire safety is a special case of safety in general and most practices which contribute to the one contribute to the other. Good housekeeping is an example, for careless storage of materials or cluttering of aisles can constitute a hazard in the event of fire and may contribute to the fire itself. In going to work in a new area for the first time it is essential to familiarise oneself with the fire warning systems and alarm points and escape routes, these normally being well signposted. Fire or smoke doors should always be kept closed since two of the most dangerous fire hazards are fumes and smoke and apart from helping to reduce the spread of the fire itself, fire doors also reduce the spread of these hazards. Certain materials, particularly certain plastics, give off toxic fumes when on fire and these should be stored away from the working area whenever possible.

If fire should break out and it is possible to do so without personal risk it should be tackled with one or other of the appliances provided.

If the room is to be vacated doors and windows should be closed before leaving, again if it is possible to do so without risk.

The fire alarm should be sounded as soon as possible and it should not be assumed that someone else has done it or will do it. If the fire alarm sounds at work even if there is no indication of a fire the fire drill for the work place should be followed. This usually entails closing all windows and doors, opening all electrical switches and vacating the premises as quickly as possible but without panic. Panic is an additional hazard and may result from being unable to see clearly because of smoke or failure of artificial lighting. Keeping a cool head is not easy in emergencies but it is essential and may be the more easily ensured if one is familiar with safety . procedures and practice.

First Aid

Almost all places of work have either professional medical assistance or persons available who are skilled in first aid and in those which do not a first aid box or kit is available. In either case, however, the first person on the scene of an accident should be aware of basic procedures. Often, particularly in the case of electric shock, delay in offering assistance may result in death. In offering general first aid at any accident the following rules are most useful and should be known.

1. Any existing hazard must be removed or if this is absolutely impossible the injured person should be removed from the hazard. The first action is to note the hazard and attempt to remove it, for example, electricity or gas must be turned off, toxic fumes or smoke dispersed, fire extinguished if this is possible. If the hazard definitely cannot be removed, the patient must be, but it should be borne in mind that movement of an injured person can aggravate or worsen the injury and extreme care must be exercised.

Removal of the hazard in the case of electric shock means turning off the electricity supply. The following points should be carefully noted.

Switch off the electricity supply if there is a continuing danger from it. This may be the case because the passage of electric current through the body activates muscles and may cause the recipient of the shock to grip the live conductor more firmly thus making the shock worse. If the supply cannot be removed, remove the recipient but *under no circumstances* use your own hands directly since the current flow may travel through to you too. Look around for a dry rope or cloth. A scarf or towel (if dry) will do. Throw this around the person and pull away from the conductor.

2. Determine whether or not the patient is conscious.

3. If the patient is unconscious check whether or not he or she is still breathing. If breathing has stopped or if the face is blue and there is breathing difficulty lift the chin and tilt the head back. If breathing does not start or improve, remove any obstruction from the mouth or throat (for example dentures) and commence mouth to mouth artificial respira-

tion as described below. If face colour does not improve or the pulse is absent cardiac massage should be employed. This is also described later. These techniques should be maintained until either help arrives or breathing is re-established. Once you are sure the patient is breathing without assistance, place him in the recovery position, lying on the side with the head sideways and the mouth clear. In case of spinal or other injury when the patient is turned to the recovery position ensure that the hips, shoulder and head turn together.

Even though breathing may be restored do not leave an unconscious person alone or lying on the back. The tongue may fall to the back of the mouth and cause suffocation.

4. If the patient is conscious determine whether or not there is bleeding. If there is, apply pressure on the wound either with or without a dressing (if something suitable is not available) unless there is glass in the wound, in which case, apply pressure at the sides of the wound. Do not tie a tourniquet around a limb or apply cotton wool or lotion to a wound. An arm or leg which is bleeding should be raised. If a dressing is not available use a *clean* handkerchief and if blood soaks through the first cover, apply a second.

5. If there is evidence of broken limbs do not attempt to straighten or bend them but wait for qualified assistance.

6. Do not give tea, coffee, brandy or any stimulant to an injured person in case hospital treatment is required. Giving drinks may delay the administering of an anaesthetic if one is needed.

7. Skin affected by irritants should be washed thoroughly but *carefully* after removing any contaminated clothing. Eyes affected by chemicals or fumes should be washed using an eyewash bottle, or, if one is not available, by putting the face into a bowl or sink of *clean* water several times. Medical aid must be sought immediately.

8. Burns caused by fire rather than chemicals should be cooled by clean running water unless the damage is obviously too severe and a sterile or clean dressing applied. Cotton wool or lotion should not be applied to burns and if anything is sticking to the burnt area it should not be removed except by qualified medical personnel.

Artificial Resuscitation

The act of breathing takes into the lungs the supply of oxygen which is vital for life to continue. The oxygen is absorbed into the bloodstream and carried to the brain and to the heart among other parts of the body. The brain and heart cannot survive in a normal state for longer than four minutes without oxygen.

After this period changes take place which are not reversible and although life within the body may continue, permanent brain damage will almost certainly ensue. It is therefore essential to get the patient breathing again as rapidly as possible and there is no time to study books or diagrams to see how it is done. The act of getting the patient to breathe again is called artificial resuscitation (or respiration) and at least one of the standard methods should be known. By far the commonest method nowadays and the easiest to apply is the *mouth to mouth* method, known usually as the 'kiss of life'.

The mouth to mouth method of artificial respiration is based on the fact that although the breath normally exhaled contains carbon dioxide and other gases not required by the body, it also still contains a proportion of oxygen which has not been absorbed. In the absence of the patient breathing for himself it is this oxygen which can maintain life until breathing is restarted. The act of breathing into the patient and allowing the breath to come out at a regular number of times per minute can also start the patient breathing on his own.

The procedure is as follows

1. Clear out the patient's mouth and throat of any obstruction, for example, dentures.

2. With the patient lying on his back, tilt the head back and keep the mouth closed.

3. Fill your own lungs, part your lips widely over the patient's mouth, pinch his nostrils together (except in the case of a child) breathe steadily into the patient and watch the chest expand. Do *not* breathe too much air into the patient.

4. Open the patient's mouth and allow his lungs to empty.

5. Repeat six times fairly quickly then continue ten times a minute (or twenty shallow breaths a minute for a child).

6. Carry on until the patient breathes again unaided. If this does not happen within a moment or two apply cardiac massage intermittently as described below.

Cardiac massage (or compression) is a means by which the heart and lungs may be restarted by external pressure on the breast bone. The position of application is most important for, in the wrong place, the technique may do more harm than good and further injure the patient.

1. With the patient laying on his back place the hands over the lower half of the breast bone (see figure A4) *but not below it* and *not in the pit of the stomach*.

2. Press down firmly to give between one and two inches of movement then release.

3. Repeat once a second for 15 seconds then give two mouth to mouth inflations. (For a child, press more lightly and repeat at a faster rate.)

4. Repeat the process until breathing starts or medical help arrives. If

breathing starts and continues unaided turn the patient to the recovery position (as described earlier) but never leave an unconscious patient after resuscitation for breathing may stop again.

A flow diagram showing the steps to be taken in deciding to apply artificial respiration techniques is given in figure A5.

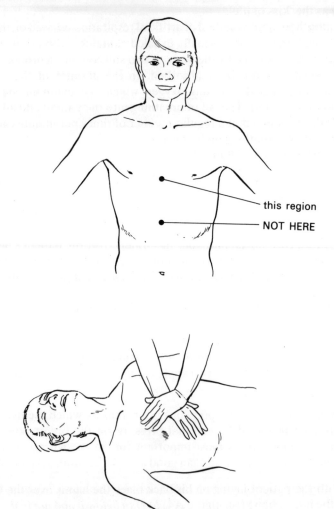

this region

NOT HERE

figure A4   Cardiac massage

To summarise this section:

    1. Think safety.

2. Get into the habit of safe working; think ahead as to how an accident might happen and try to avoid the possibility.

3. Remember good housekeeping.

4. Learn about fire alarms, routes and extinguishers in your place of work.

5. Learn the rudiments of first aid, the mouth to mouth method of artificial respiration and the basic technique of cardiac massage.

6. If an accident happens try not to panic, do not waste time thinking—act.

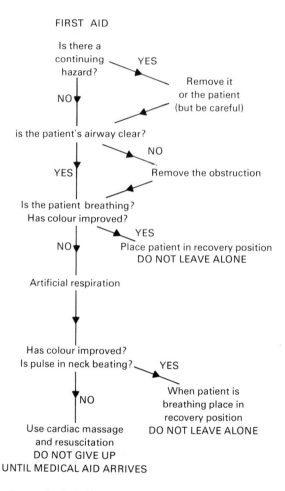

FIRST AID

Is there a continuing hazard?   YES
Remove it or the patient (but be careful)

NO

is the patient's airway clear?   NO
Remove the obstruction

YES

Is the patient breathing?
Has colour improved?   YES
Place patient in recovery position
DO NOT LEAVE ALONE

NO

Artificial respiration

Has colour improved?
Is pulse in neck beating?   YES
When patient is breathing place in recovery position
DO NOT LEAVE ALONE

NO

Use cardiac massage and resuscitation
DO NOT GIVE UP
UNTIL MEDICAL AID ARRIVES

figure A5   Flow diagram for first aid

Appendix 2: Resistor colour code and schematic diagram code BS 1852

The resistance of most fixed resistors (other than wirewound types) is usually indicated by the use of a system of coloured rings grouped together at one end of the component. There are usually either three or four of these rings and they are read as follows: rings one and two, counting from the end at which the rings are grouped, give the first two figures of the resistance value; ring three gives a number of noughts following these two figures; and ring four (if present) gives the tolerance (that is, a measure of the range within which the actual resistance is permitted to lie). The colour code (table A1) is as follows:

Colour	Black	Brown	Red	Orange	Yellow	Green
Figure (or number of noughts, for · third ring)	0	1	2	3	4	5

Colour	Blue	Violet	Grey	White	Silver	Gold
Figure (or number of noughts, for third ring)	6	7	8	9	—	—

table A1

If a fourth ring is present it will be silver to indicate 10% tolerance or gold to indicate 5% tolerance. If a fourth ring is not present the tolerance may be assumed to be 20%.

Thus, a resistor coloured red-red-red-silver has a nominal resistance of 2200 (that is, 2–2–two noughts) ohms and its actual resistance lies within ± 10% of 2200, that is between (2200 − 220) ohms and (2200 + 220)

262

ohms. Similarly, a resistor coloured yellow-violet-brown is 470 ohms, 20% tolerance. A third ring coloured black signifies no nought so that a brown-black-black set of rings would indicate 10 ohms, that is first two figures 1 and 0, no noughts to follow.

BS 1852 Resistance Code

Throughout the book schematic circuit diagrams have indicated resistance values using standard prefixes. The new British Standard 1852 Code is being adopted by some manufacturers. This code tells more about the resistor but uses fewer characters. Some examples are

$0.56\ \Omega$ is written R56
$1.0\ \Omega$ is written 1R0
$5.6\ \Omega$ is written 5R6
$68\ \Omega$ is written 68R
$2.2\ k\Omega$ is written 2K2
$10\ M\Omega$ is written 10M

An additional letter may then be used to give the tolerance.

$F = \pm 1\%$ $J = \pm 5\%$
$G = \pm 2\%$ $K = \pm 10\%$ $M = \pm 20\%$

So that

4R7K means $4.7\ \Omega \pm 10\%$
68KK means $68\ k\Omega \pm 10\%$
4m7M means $4.7\ M\Omega \pm 20\%$
6K8J means $6.8\ k\Omega \pm 5\%$

and so on.

Self-test questions and answers

Questions

1. The type of battery used in vehicles for starting purposes is usually

A. Lead–acid
B. Leclanché
C. Alkaline–manganese
D. Mercury

2. In the mercury cell the anode is made of

A. Mercury
B. Mercuric oxide
C. Zinc
D. Potassium hydroxide

3. The depolariser in an alkaline–manganese cell is mainly

A. Potassium hydroxide
B. Manganese dioxide
C. Powdered zinc
D. Caustic potash

4. The unloaded e.m.f. of an unused Leclanché cell is

A. 1.0 V
B. 1.2 V
C. 1.4 V
D. 1.6 V

5. In a fully charged lead–acid cell the positive active material is

A. Lead sulphate
B. Spongy lead
C. Lead peroxide
D. Sulphuric acid

6. The negative active material in a discharged lead–acid cell is

A. Lead sulphate
B. Spongy lead
C. Sulphuric acid
D. Lead peroxide

7. The negative electrode of an alkaline cell contains

A. Iron or cadmium
B. Nickel
C. Zinc
D. Potassium hydroxide

8. A lead–acid battery is able to provide a current of 25 A for 10 h. The capacity at the 10-h rate for this battery (in Ah) is

A. 2.5
B. 250
C. 0.4
D. Not determinable without further information

9. The colour of both plates in a discharged lead–acid cell is

A. Black
B. Whitish grey
C. Brown
D. Slate grey

10. The specific gravity of an alkaline cell when discharged is 1.19. When fully charged the specific gravity is

A. Lower
B. Higher
C. Same
D. Not determinable without further information

11. The unit of electric charge is the

A. Coulomb

B. Volt
C. Ohm
D. Ampere

12. The unit of electric current is the

A. Coulomb/second
B. Volt
C. Ohm
D. Coulomb

13. The unit of electrical resistance is the

A. Siemens
B. Ampere
C. Ohm
D. Volt

14. The total equivalent resistance of two resistors of resistance R_1 and R_2 in series is equal to

A. $\dfrac{1}{R_1} + \dfrac{1}{R_2}$

B. $\dfrac{R_1 R_2}{R_1 + R_2}$

C. $R_1 + R_2$

D. $\dfrac{1}{R_1 + R_2}$

15. The total equivalent resistance of two resistors of resistance R_1 and R_2 in parallel is equal to

A. $\dfrac{1}{R_1} + \dfrac{1}{R_2}$

B. $\dfrac{R_1 R_2}{R_1 + R_2}$

C. $R_1 + R_2$

D. $\dfrac{1}{R_1 + R_2}$

16. The resistance of a piece of material of resistivity ρ, length l and cross-sectional area a (all in SI units) is given by

A. $\dfrac{1}{\rho a}$

B. $\dfrac{\rho a}{l}$

C. $\dfrac{\rho l}{a}$

D. $\dfrac{a}{\rho l}$

17. A piece of material has an electrical resistance of 10 Ω. If its length is doubled and its cross-sectional area halved the new resistance is

A. 10 Ω
B. 40 Ω
C. 2.5 Ω
D. Incalculable without more information

18. The resistance of a piece of material at a temperature $t\,^{\circ}$C in terms of its resistance at 0 $^{\circ}$C, R_0 Ω and the temperature coefficient of resistance α is given by

A. $1 + \alpha t$
B. $R_0 + \alpha t$
C. $R_0 + \alpha R_0 t$
D. $\alpha R_0 t$

19. The resistance R_2 of a material at any temperature t_2 in terms of its resistance R_1 at any other temperature t_1 and the temperature coefficient of resistance α is given by

A. $\dfrac{R_2}{R_1} = \dfrac{1 + \alpha t_1}{1 + \alpha t_2}$

B. $\dfrac{R_2}{R_1} = \dfrac{1 + \alpha t_2}{1 + \alpha t_1}$

C. $R_2 = R_1(1 + \alpha t_1)(1 + \alpha t_2)$

D. $R_2 = R_1 + \alpha t_2 - \alpha t_1$

20. Power in a d.c. circuit in terms of voltage V, current I and resistance R is given by

A. *IR*
B. *V/R*
C. *VI*
D. $\dfrac{VI}{R}$

21. The minimum power rating of a 10-Ω resistor connected across a 100-V supply is

A. 10 W
B. 100 W
C. 1 W
D. 1 kW

22. The unit of magnetomotive force is the

A. Tesla
B. Weber
C. Ampere-turn
D. Henry

23. The unit of magnetic flux is the

A. Tesla
B. Weber
C. Ampere-turn
D. Henry

24. The unit of magnetic flux density is the

A. Weber
B. Ampere-turn
C. Tesla
D. Henry

25. The unit of the coefficient of self-inductance is the

A. Henry
B. Weber
C. Tesla
D. Ampere-turn

26. The current flowing in a wire placed in a magnetic field of flux density 0.2 T when the force per metre length of the conductor is 0.6 N is equal to

A. 3 A

B. 0.12 A
C. 0.33 A
D. Incalculable without further information

27. The voltage induced per turn of a coil subjected to a changing magnetic field is equal to

A. Flux/time
B. Current/time
C. Rate of change of flux with time
D. Rate of change of flux with current

28. The voltage induced across a coil in terms of its coefficient of self-inductance L is equal to

A. $L \times$ flux/time
B. $L \times$ current/time
C. $L \times$ rate of change of current with time
D. $L \times$ rate of change of flux with time

29. The back e.m.f. induced across a coil of self-inductance 0.4 H when a 1.5 A current is reversed in 0.2 s has an average value of

A. 3 V
B. 6 V
C. 53.3 mV
D. 26.67 mV

30. The e.m.f. across a coil is 4 V when the current through an adjacent coil is changing at the rate of 2 A/s. The mutual inductance between the coils is

A. 2 H
B. 8 H
C. 0.5 H
D. 1 H

31. The periodic time of a 60-Hz sine wave is (in seconds)

A. 60
B. 0.0167
C. 16.67
D. 0.00833

32. If a wave enters a medium normally

A. Its velocity always remains unchanged

B. Its direction remains unchanged
C. Its velocity always increases
D. Its direction always changes

33. The image formed by a plane mirror is

A. Virtual and inverted
B. Real and inverted
C. Virtual and erect
D. Real and erect

34. For a biconvex lens, if the object is at a distance along the principal
axis greater than twice the focal length, the image is

A. Real, inverted and larger
B. Real, erect and larger
C. Virtual, inverted and diminished
D. Real, erect and diminished

35. For a biconvex lens, if the object is at a distance along the principal
axis equal to twice the focal length, the image is

A. Real, erect, same size
B. Real, inverted, same size
C. Virtual, erect, larger
D. Real, inverted, smaller

36. For a biconvex lens, if the object is at a distance along the principal
axis less than twice the focal length but greater than the focal length, the
image is

A. Virtual, inverted, diminished.
B. Real, erect, same size
C. Real, inverted, magnified
D. Virtual, inverted, magnified

37. The periodic time of a wave of frequency 50 Hz is

A. 50 s
B. 20 ms
C. 0.2 s
D. 50 cycles/s

38. If the propagation velocity of an electromagnetic wave is 3×10^8 m/s,
the wavelength of such a wave having a frequency of 1 MHz is

A. 300 m

B. 3×10^{14} m
C. 0.0033 m
D. 150 m

39. For a wave of constant propagation velocity

A. The higher the frequency the longer the wavelength
B. The lower the frequency the shorter the wavelength
C. The higher the frequency the shorter the wavelength
D. The frequency and wavelength are constant

40. Which of the following statements is true?

A. Light waves have a higher frequency that X-rays
B. X-rays have a shorter wavelength than cosmic rays
C. Ultraviolet rays have a higher frequency than gamma rays
D. Gamma rays have a higher frequency than X-rays

41. The main disadvantage of d.c. signal transmission is

A. Only low power signals may be transmitted
B. d.c. power supplies are required
C. Electromagnetic propagation cannot be used
D. The extent of the transmitted intelligence is limited

42. The synchronising signal in a television transmission is used to

A. Separate sound from vision signals
B. Ensure line and frame timebases run at the same frequency
C. Ensure camera and receive timebases run in synchronism
D. Ensure the same brightness at the camera and at the receiver

43. A superheterodyne receiver system

A. Has poor selectivity
B. Uses a frequency changer
C. Has poor sensitivity
D. Uses wide band amplifiers

44. An AND gate gives an output 1 if

A. At least one input is 1
B. At least one input is 0
C. All inputs arc 0
D. All inputs are 1

45. With all inputs at 1

A. An AND gate gives an output 0
B. An inclusive OR gate gives an output 0
C. A NAND gate gives an output 1
D. A NOR gate gives an output 0

46. Two NOR gates connected in cascade give an equivalent function to one

A. NOR gate
B. AND gate
C. OR gate
D. NAND gate

47. An OR gate gives an output of 0 if

A. At least one input is 1
B. At least one input is 0
C. All inputs are 1
D. All inputs are 0

48. A two input logic gate produces an output 0 when both inputs are 1. The gate could be

A. An OR gate
B. An AND gate
C. An inverter
D. An exclusive OR gate

49. In a three-element Yagi aerial with one director and one reflector

A. All elements are the same length
B. The director is longer than the reflector
C. The director is shorter than the reflector
D. The director and reflector are of equal length and shorter than the dipole element

50. Two NAND gates connected in cascade give an equivalent function to one

A. NOR gate
B. AND gate
C. OR gate
D. NAND gate

Answers

1.	A	26.	A
2.	C	27.	C
3.	B	28.	C
4.	D	29.	B
5.	C	30.	A
6.	A	31.	B
7.	A	32.	B
8.	B	33.	C
9.	A	34.	A
10.	C	35.	B
11.	A	36.	C
12.	A	37.	B
13.	C	38.	A
14.	C	39.	C
15.	B	40.	D
16.	C	41.	C
17.	B	42.	C
18.	C	43.	B
19.	B	44.	D
20.	C	45.	D
21.	D	46.	C
22.	C	47.	D
23.	B	48.	D
24.	C	49.	C
25.	A	50.	B

Index